The Jamaica Planter's Guide

*Or, a System for Planting and Managing
a Sugar Estate or Other Plantations
in that Island, and Throughout the
British West Indies in General*

THOMAS ROUGHLEY

CAMBRIDGE
UNIVERSITY PRESS

CAMBRIDGE UNIVERSITY PRESS

Cambridge, New York, Melbourne, Madrid, Cape Town, Singapore,
São Paolo, Delhi, Dubai, Tokyo, Mexico City

Published in the United States of America by Cambridge University Press, New York

www.cambridge.org
Information on this title: www.cambridge.org/9781108024303

© in this compilation Cambridge University Press 2010

This edition first published 1823
This digitally printed version 2010

ISBN 978-1-108-02430-3 Paperback

CAMBRIDGE LIBRARY COLLECTION

Books of enduring scholarly value

History

The books reissued in this series include accounts of historical events and movements by eye-witnesses and contemporaries, as well as landmark studies that assembled significant source materials or developed new historiographical methods. The series includes work in social, political and military history on a wide range of periods and regions, giving modern scholars ready access to influential publications of the past.

The Jamaica Planter's Guide

Thomas Roughley (fl. 1820) published this *Guide*, based on his own experience, in 1823. It is an important source on Britain's richest colony, where the sugar trade had reached its peak around 1810. At a time when the abolition of slavery was a major issue, Jamaican planters were particularly concerned, as so much of their activity was based on slave labour. The book deals with all aspects of running a sugar plantation profitably, the chapter on the work force being particularly interesting. He outlines each role necessary for the effective estate, and criticises the 'slanders' of philanthropists regarding the mistreatment of slaves, referring to the strict laws for their protection: since the importation of new slaves had been banned since 1807, it was all the more necessary for owners to look after their workforce's health and welfare. Roughley shows that well-run plantations were highly complex agricultural and economic units.

Cambridge University Press has long been a pioneer in the reissuing of out-of-print titles from its own backlist, producing digital reprints of books that are still sought after by scholars and students but could not be reprinted economically using traditional technology. The Cambridge Library Collection extends this activity to a wider range of books which are still of importance to researchers and professionals, either for the source material they contain, or as landmarks in the history of their academic discipline.

Drawing from the world-renowned collections in the Cambridge University Library, and guided by the advice of experts in each subject area, Cambridge University Press is using state-of-the-art scanning machines in its own Printing House to capture the content of each book selected for inclusion. The files are processed to give a consistently clear, crisp image, and the books finished to the high quality standard for which the Press is recognised around the world. The latest print-on-demand technology ensures that the books will remain available indefinitely, and that orders for single or multiple copies can quickly be supplied.

The Cambridge Library Collection will bring back to life books of enduring scholarly value (including out-of-copyright works originally issued by other publishers) across a wide range of disciplines in the humanities and social sciences and in science and technology.

THE

JAMAICA

PLANTER'S GUIDE.

LONDON:
Printed by A. & R. Spottiswoode.
New-Street-Square.

THE

JAMAICA

PLANTER'S GUIDE;

OR,

A SYSTEM FOR PLANTING AND MANAGING

A

SUGAR ESTATE,

OR OTHER PLANTATIONS IN THAT ISLAND,

AND THROUGHOUT

THE BRITISH WEST INDIES

IN GENERAL.

ILLUSTRATED WITH INTERESTING ANECDOTES.

———

By THOMAS ROUGHLEY,

NEARLY TWENTY YEARS A SUGAR PLANTER IN JAMAICA.

———

LONDON:

PRINTED FOR

LONGMAN, HURST, REES, ORME, AND BROWN,

PATERNOSTER-ROW.

1823.

Academiæ Cantabrigiensis
Liber.

PREFACE.

At any period, but more especially the present, when Jamaica and West India produce is so much depreciated, and the heavy expenses attending the cultivation of the land, and the manufacture of its produce; together with the capital laid out in the establishment of a plantation, and its annual disbursements, is considered, a system of practical cultivation, combined with proper economy, will, it is presumed, be readily adopted by every proprietor. An attention to such a plan will doubtless appear necessary in all cases of the non-residence of the owners, and where the management of estates is intrusted to agents. Many circumstances may concur to render such care indispensable — the failure of crops, the unusual decrease and

loss of slaves and stock, the exhaustion of
land, inadequate and trifling returns by
shipments or sales, a consequent diminution
of capital, and the inevitable alternative
(after many years of toil and anxious hope)
of borrowing money by mortgaging his
estate, to sustain his credit. To these
contingencies may be added law charges,
interest of money, and per centage to
agents ; and at last a consignment of the
produce to a mortgagee, or putting him in
possession of the estate, thus perhaps blight·
ing his prospects for ever.

The author of this work has spent many
years in the Island of Jamaica, in the occupation of a planter, and in the management
of several estates, principally in the north
side of that productive, delightful island.
His knowledge of the prevailing system of
culture has been matured by experience,
and he has, he trusts, discovered some of
those errors which have occasioned both
expense and failure. He has therefore
ventured to arrange such a method of

plantership (especially for the sugar department) as is best adapted, not for Jamaica alone, but for the West India islands in general. He has introduced some useful, comprehensive, and beneficial modifications, which he doubts not will be conducive to the interests of the proprietors. He presumes that such a plan will meet with the approbation, countenance, and support of the independent, non-resident owners of land, and capitalists, and all who are interested in the welfare of that species of property. He has endeavoured to make it simple and of easy acquisition, particularly so to the beginner; yet, he trusts, of equal value to the old practitioner. In the work he has also inserted some interesting anecdotes founded on facts, which have either fallen under his own observation, or been received from credible testimony. He trusts that his work will be found by the impartial and unbiassed of general utility, and that it will therefore meet with approbation.

CONTENTS.

JAMAICA

PLANTER'S GUIDE.

CHAP. I.

OF PLANTATION ATTORNEYS OR AGENTS.

In treating of the management and cultivation of a sugar estate in the island of
Jamaica, it will be necessary to take into
consideration a great variety of subjects,
and to reduce them into such a system as
will enable the speculator to approve and
adopt it, and the practitioner to understand
and realise it. The plan here offered to
public attention is intended to unite proper
economy in every branch, with such requisites as may ensure (under the blessing
of Providence) a suitable return from the
land, and yield the proprietor a reasonable
and adequate compensation for the employ-

B

ment of his capital. The first and principal object to be contemplated is, the course which is at present followed in the island of Jamaica, with respect to the treatment of slaves and stock, the culture of the land, the manufacture of sugar and rum, and the necessary and unavoidable expenses connected with its produce. An accurate view of these subjects will determine, whether improvement and retrenchment may generally and successfully be attempted. If this should be found the result of the investigation, an increase of the crops, the preservation of the capital, the improvement of the quality and quantity of sugar and rum, may be obtained; the comfort of the slaves, the food and care of the live stock, and the good order of the buildings and manufacturing utensils may be secured; the shipment of the produce in proper condition, with little reduction of value for repacking or ullage, and the disposal of the whole in as speedy a manner as possible, to the advantage and satisfaction of the proprietor, without diminishing the salary or happiness of the white people resident on the estate, may be expected.

The author of this treatise disclaims all idea of personal reflection, though it is possible that some may invidiously conjecture this to have been his intention. If, indeed, by any recital he has made, it should appear that any abuses are found to exist, the representation is given as growing out of a system of management which has for many years been adopted in Jamaica — a system tenaciously retained by the old planters; patronised by the new; and become so formidable by long use and custom, as scarcely to admit of innovation or improvement. Useful modification too often, therefore, is condemned with sarcasm and ridicule.

The person styled in Jamaica, and the West Indies in general, by the planters, an Attorney, is so important an object in the detail of plantership, that I should be grossly deficient in describing the old customs and habits of those islands, if some account was not given of him. These officers were first appointed there in consequence of the absence of the proprietors. This malady became so epidemical as to affect and carry off from Jamaica seven eighths of the landed interest, substituting in their place, during their absence, some gentleman of their acquaintance, or a person recommended to them, to conduct the business of the plantation. The enjoyments which the mother country affords, its elegancies and refinements, frequently became so alluring, as to induce a protracted continuance in it. Their affairs in the West Indies were, therefore, necessarily confided to the management of an agent or attorney, resident on or near the estate, who acted according to his own discretion and pleasure. Some gentlemen of this class possessed only a slight knowledge of plantership, and others none at all. But being armed with despotic power, they became the ruling agents in the affairs they were intrusted with ; and enjoying an extensive connection with mercantile men in the islands in general, and the parent country, they were looked up to with reverential awe by the overseers and book-keepers. Sometimes they paid a casual visit to the estates over which they presided, with lordly pomp. This, however, was seldom done oftener than once a-year, and the plantations were governed without being actually inspected. Persons of merit were frequently discharged from their stations without sufficient cause, and others employed instead of them, of whose qualifications they were ignorant. Plans were formed, from the adoption of which, prolific crops were anticipated. But the fairest prospects were sometimes made to vanish by an arbitrary stroke of the pen, so that happy results and promised returns yielded to the destructive consequences of ill-concerted schemes. Such conduct was more the effect of malignant caprice, than the solicitous effects of honourable intentions, for promoting the welfare and interest of the proprietors. These were the primary causes of the failure of the crops, and the springs from which all the subsequent evils flowed, — " the very head and front of the offending." This occasioned the spreading the cane cultivation over an immense tract of land, instead of keeping up the stamina of the old grounds. The cattle and mules were overworked, or suffered to grow too old without being replaced by younger beasts. Hence a great loss was sustained : the introduction of endless jobbing : the eating up of the crops by divers island disbursements, which might have been avoided or diminished : and the immense supplies sent out by home merchants, a great part of which were left to rot, or were buried at the wharf. The sugar and rum were suffered to remain too long on the estate or barquadier without shipping, requiring sometimes one quarter to make up loss by ullage. These gross errors are solely the fault of the resident attorney, arising from his connivance or ignorance. The consequences, however, of this misconduct, and of these improprieties, fall on the owner of the estate, obliging him to borrow money, not only for his own support, but to pay off the debts of his ill-managed and unproductive land. I would not indeed assert, that in the island of Jamaica in particular, no gentlemen are to be found, properly qualified by nature or education, to render, as planting attorneys, the most excellent, durable, and beneficial services to those who employ them. There are many who are fully adequate to discharge such a trust. Governed by honourable motives, and assiduously employing their cultivated talents, by persevering industry, they secure the welfare of their constituents; and actuated by humane and liberal principles, they extend the advantages they obtain, to all who are dependent upon them. The major part, or five sixths of this class of representatives, are men who are engrossed by their own interested speculations, attentive to what will promote their own selfish views, making every other object within their grasp subservient to their ambition. Though perhaps they are too ignorant to be planters, they are too ostentatious, proud, and supine, to contribute to the good of their constituents. Nor is it the least of their defects that they clandestinely prefer the interest of a mortgagee or affluent merchant, (who may possibly have some secret design upon the estate) to that of the owner. This is sometimes done, because such a mortgagee or merchant, becoming powerfully connected in the island, may be the means not only of continuing the attorney in his agency, but of advancing his influence, authority, and wealth, by procuring for him commissions for the management of other estates. By this means, a fruitful West India island is converted into a devastated mortgage. Sometimes it happens that the attorney, consulting his pleasure, fixes his residence on one of the estates of his constituents, with far greater regard to

save expense to himself, than his absent
employer. He looks about for a healthy
eligible situation; a large house well furnished; sensible, clean, handsome slaves
as servants; grounds delightful and pleasing
to his eye; a flower garden to refresh his
weary senses; a kitchen garden to furnish
his table; but, above all, where small stock
are numerous and thriving, that he may
have an abundant supply for his own consumption without expense. He cultivates
luxuriant grass pieces, and corn for his
horses and cattle, without any deduction
from his per centage on the sale of the
crops. Thus his establishment is secretly,
and with mercenary meanness, supported:
the proprietor of the estate in this manner
contributing to his aggrandisement. The
contribution thus levied, is not less, considering all circumstances, than 700l. a
year. What, however, is still more surpris-
ing, is, that this very person, who thus
reposes upon the substance of his constituents, in superabundant plenty, very

often does not know the limits of the estate he is empowered to manage, or even a cane piece within a musket shot of the house he inhabits.   He cannot tell whether it is attached to that or any other property. The poor overseer, dependent on his will, is in constant jeopardy, and his mind frequently thereby estranged from the proper duties of his station.   If any deficiency is discovered in the supplies required for the house or retinue of the attorney, he is reproved by harsh notes, and suppositious faults are imputed to him.   Whether an ample crop has been taken off, and a promising one is advancing, with little loss of slaves and stock (the true criterion of good management) signifies little.   This meritorious, industrious servant, is discharged, and his place filled up with perhaps an obsequious novice.

Some proprietors of West India estates are so profuse, or rather overcareful of their property, as to suppose that their advantage is not sufficiently attended to,

unless they have two guardians on the spot;
one termed the planting attorney, the other
the commissary, factor, or mercantile attorney. Each of these receives a commission of five per cent. on all sales or net
proceeds. These two persons frequently
pursue a separate interest, and clash with
each other. They profess to discover
blunders or abuses in their different de-
partments, and so greatly confuse the
general affairs of the estate, by their mutual
variance and opposition, that the ordinary
and requisite materials and business of the
plantation is neglected. Their personal
interest being separate, they conspire to
supplant each other, at the risk or even
ruin of the proprietor. It often happens
that the power of these men is so pe-
culiarly vested, that under particular circumstances they can appoint each other
to office: either the planting attorney appointing the factor, or, on the contrary, the
factor nominating the planting agent. In
such cases, the disease is equally mortal to

the owner. For one of these persons predominating over the other, of course, orders, dictates to, and rules the other. The merchant will enlarge the prospects, and contribute to the interest of the planting agent, from motives of gratitude to his benefactor, the planting agent; he will ship round and consign to the factor, a great part of the produce of the estate, instead of sending it home. In the place of this the superabundant supplies, and a quantity of worthless lumber, and salt provisions, are transmitted back. This abuse of the trust reposed in them is ruinous to the estate. By this means, a great part of the crop becomes as it were sequestered; destruction is added to defalcation. The property which is disposed of in this manner is converted into bills of exchange. These are sold at a high premium, though they virtually and intrinsically belong to the land owner. By this twofold stratagem, designed to fill the coffers of the agents, a great portion of the produce of

Jamaica is sold in Kingston, either by private contract or by auction, in as summary
a manner as possible, to make quick returns to the factor. By these methods he
becomes the proprietor of it at a cheap
rate, which is indeed the object he had in
view throughout the whole of his transactions.

Some short-sighted people may, perhaps,
indulge the idea that it is contrary to the
interest of the acting attorney to be influenced by principles, and to pursue measures contrary and detrimental to his
constituent. I will indeed allow that honourable motives, and views characterised
by integrity, should predominate with every
man so entrusted, and whose connections
enable him to promote the welfare of his
employer: and that if the agent is ignorant
of plantership, he should decline the office;
or if he feels himself competent to the task,
he should discharge it with faithful assiduity: but whoever forms this honourable view of the subject has seldom, or

possibly never enquired into the cause of the absence of the proprietor, or traced the source from whence the power of the attorney is derived, nor has he properly considered the sordid and selfish views of these agents. When a mortgagee finds an unfortunate proprietor embarrassed and in trammels, he has immediately an interest of his own to secure, and appoints an agent who will be subservient to his purpose. A letter, enigmatically expressed, but well understood by the resident agent, is despatched to him. The list of supplies is augmented ; the ordinary expenses are increased to double or treble the former amount ; the interest of borrowed money accumulates ; the crops are diminished ; shipments are delayed, by which great loss is sustained, though affectedly deplored ; an overseer, indifferent as to events or results, not only suffered to remain on the estate to manage it, but frequently encouraged in his misconduct ; the mortgage is foreclosed, or a decree in chancery issued

for the sale of the estate, which is disposed
of for a trifling consideration to some person employed to purchase it for the mortgagee or attorney. Such circumstances
account, in part, at least, for the conduct
of some of the planting attorneys in not
consulting, as they ought, the advantage of
those who employ them, and who are
originally the owners of the land. Having
thus gained the confidence of the mortgagee, by betraying the interest of the
proprietor, and rendered the former this
service of spoliation, he obtains powers to
act for others, and both his celebrity and
his fortune are established. In confirmation
of this, I shall here relate an instance
which occurred some years ago in Jamaica.
In the mountainous parts of the parish of
Clarendon there was an estate much diversified by hills. Though apparently unpro-
mising, yet by laying out and making
roads of easy ascent, at a great expense
indeed, the proprietor found he could convert it into a valuable plantation, it being

of great extent and the quality of the land good. A convenient place was chosen on which to erect a set of works. These were constructed upon a large scale. A numerous gang of effective negroes, and an adequate quantity of cattle and mules were provided. The estate yielded great crops. Six hundred hogsheads of sugar, of good quality, were obtained from it. But it so happened that in the commencement of these operations a sum of money was borrowed to forward the undertaking. Large supplies of heavy articles became necessary; this, with accumulating interest on the money at first borrowed, soon created embarrassment. The merchant who had furnished the supply of money and goods became impatient for payment, before the crops could be disposed of to reimburse him. He knew there was a large capital that could be laid hold of, and therefore commenced a suit in chancery against the property. This is the mode generally adopted to wrest the possession of land from the owners in this island. The accounts on both sides were obliged to undergo the revision of a Master in Chancery. Large deductions were made from the value of the estate, by the expenses of the suit in question. Years rolled on in this civil warfare, constantly swelling the account. Every year fresh examinations were made for the Master in Chancery. The decree was at last given in favour of the plaintiff, allowing him all his costs. During the whole of the time a receiver was appointed by the Master in Chancery. In such cases it has happened that the plaintiff was on friendly terms with the Master, in consequence of which the original overseer has been exchanged for one more subservient to the views of the plaintiff. The supplies are not diminished, though the estate is declining in its produce. The works are suffered to fall into decay. The negroes assume a poor, weakly appearance. Every thing sinks in value. Preparatory to the sale of the estate in question, an estimate

c

was made by appraisement of the negroes, the stock, the land, and the works. This was produced on the day of sale, and the estate was disposed of, with every thing belonging to it, for half or one third of its real price. The amount was just as much as would pay the plaintiff his demand and costs of suit. Thus a final period was put to the fortune and hopes of the unfortunate proprietor. Thus was realized the scheme of the new owner, who was well acquainted with the real value of the estate, and the crisis of its doom. He was desirous of calling it his own, though at the expense of every thing honourable and just. An illicit trafficker from a distant northern climate, he seized with avidity the lucky moment for the enterprize, no one being present who could advance so large a sum as the purchase-money but himself. He had a trusty agent present on the day of sale, to prevent the possibility of losing such a bargain, and to render the purchase for himself certain. In a short time, however, after this event,

the disposition of every thing on the
estate was altered. The negroes, the stock,
and works were restored to their former
good state and appearance, at a trifling expense; and a steady annual crop of from
five to seven hundred hogsheads of sugar
annually, were obtained from the land thus
unjustly transferred from its first rightful
proprietor, to the oppressing hand of the
artful and enterprising mortgagee.

I now come to consider this subject in
another point of view. When the proprietor
has little or no knowledge of the agent he
appoints, further than by recommendation,
perhaps seldom or ever visiting the island
or his estate, he understands little of the resources it possesses, or the culture it requires.
He does not know whether the person whom
he deputed ever was on the plantation or
even that part of the island; whether he
had either seen or knew how to plant and
bring a cane to maturity. But unwilling to
risk a voyage at sea, still less to hazard the
danger of tropical sickness, he rests satis-

fied for some time, that the affairs of his estate are going on well, under the wise, provident care of his planting attorney. The arrival, after the lapse of a few months, or the termination of a crop or two, of the sad unpleasant news, that there is a deficiency of produce, and the formal crop account, announcing a defalcation of one third of what was naturally expected, undeceives him. At the same time, he is informed, in order to make up such defalcation in future, that there is an absolute necessity to put in such a quantity of land, with the aid of a jobber. He must pay such jobber in rum at cash price : other contingencies demand the produce of the rum crop to defray them. Thus commences a train of disbursements, having for their foundation, fabricated evils and false representations, solely for the purpose of shielding from the proprietor's knowledge the gross fraud of bad and dilapidating management. Such planting attorneys as these generally have independent fortunes in the island. Careless of such events, yet greedy of gain, they continue in power for a crop or two, satisfied with their being paid their per centage, and regardless whether the estate thrives, which they have thrown back for years as to adequate returns. A new agent is appointed, a new overseer put in charge, fresh plans are adopted; and this independent attorney, after having sown the seeds of destruction, enquires at some future period how such an estate goes on, and being informed of its miserable returns, vaunts, that if it had been left to his care and management, the result would have been otherwise. Many of these men have a curious satirical vein running through their dispositions. For if their successors, with the persons whom they employ, do their duty, bring the estate from a wretched, ruined condition to prosperous thriving cultivation, regularly and steadily extending its produce, instead of the former pernicious confusion; if they procure orderly respectful demeanour in the slaves,

instead of idleness and arrogance; if their
appearance is decent and becoming, exhibiting in their persons comfort and cleanliness, instead of disgusting filthiness, disease and aversion to their proper occupations; if the cattle is preserved by good
feeding, with few sores, or deep, morbid,
putrid distempers; if the subordinate white
people are attentive to their business,
steady, sober, and quietly disposed, making
the business and welfare of the estate their
pleasure, instead of riotously galloping on
horseback about the country, imbibing and
entailing on themselves by excess the sure
consequences of early distempers, which
generally prove fatal; this exhibition of
what is right and proper, instead of being
treated with panegyric, is made, through
every medium of ridicule, envy, and detrac-
tion, to appear in an invidious, morose,
cruel, coercive, and inhospitable point of
view.   It would be in vain that such people strove to gain their favour: in vain
would they solicit employment from them:

they would be left contemned and neglected, while the profligate, vicious, abandoned, and worthless, would find employment and support. Such men indulge their
desires, lead an idle and dissipated life, and
leave the essential business of the estate, to
the venal management of head people,
whom they satisfy with an extra weekly
allowance from the rum and provision store.
But what gives such overseers a stronger
claim to the patronage of their employer
or attorney, is the loan of whatever money
they may have saved, either from their
salaries or as owners of jobbing gangs.
This hides a multitude of faults; this
smooths the road to preferment; this rivets
the lasting good opinion of the patron;
changes his vices into virtues; prevents
every effort for the preservation of the
estate ; converts the cane and grass pieces
into ruined heaps; the well disposed negroes
into runaways ; the cattle into carrion ; the
works and manufacturing houses into

wretched, unroofed walls, and the proprietor into an insolvent.

An attorney of this description will
sometimes live in some remote parish in
the island, seldom visiting the estate he is
commissioned to manage. This he does
indeed when he formally takes possession
of it. Every communication is carried on
between him and the overseer by letter, or
expresses by the negroes, on estate mules;
or else the overseer is ordered to ride over
to him, perhaps seventy or an hundred
miles, to arrange accounts and future plans.
This is a pleasurable excursion for the overseer, sometimes, and is made to last a
fortnight. The natural inclination of man
is alive to pleasure, to aversion of control,
to hatred of minute enquiry into particulars
which may be pregnant with error or vice.
to dislike of restraint or control. Such an
employer is therefore much approved of by
a majority of people throughout the West
India Islands, as giving little trouble or
disturbance by his presence, or dictating

orders. Whatever is required is granted; whatever is neglected is overlooked. The overseer becomes absolute and prosperous in his affairs. He employs and discharges his book-keepers at pleasure; gives lucrative jobs to favourite tradesmen, and in a short time acquires the name of a generous, good man — one to be looked up to; in short, he obtains the character of a good planter, although the head driver manages every thing on the estate. I have known one of these overseers, who lived more than six years upon an excellent estate in the parish of St. Mary's, in Jamaica, who did not understand the boundaries of it; and when the proprietor unexpectedly came, and expressed a wish to see and go over the lines of it, was obliged to declare his total ignorance of its limits, as he never had seen them. These the proprietor himself went to and showed him, although he had been absent near twenty years from the estate. The land had been despoiled of its valuable wood by the neighbours, and great part taken up and cultivated in negro provisions, by the slaves of other plantations, in consequence of the supineness of the acting island agent.

It may be argued, that scarcely any remedy can be found, or applied, to check this established evil, or root it out; that the disease is too inveterate to admit of cure; that numberless obstacles stand in the way of the completion of so happy, so salutary a design. But it requires no great penetration to discover the true nature and strength of this disease, this cause of injury and ruin. The sources of the malady may not only be easily discovered, but when they are once traced and pointed out, will be found so iniquitous as to shrink from publicity and exposure. With firm resolution, therefore, the bold, the decisive effort must be made to cut away not merely the branches, but to destroy the stem and root of this poisonous tree, this source of all the evils which exist. Dismiss from their employment these corrupt agents; let no sycophant advise the land owners to retain them. The estate, with prudent attention, being thus liberated from this baneful control, will soon assume the appearance of health, animation, and prosperity. Industry, with merit as its attendant, will insure success ; simple, yet energetic measures will easily attain whatever is necessary or desirable ; and having brought things to this crisis, this sharp, but sure remedy will prove effectual.

I shall now beg leave to point out the system to be acted upon, which, I presume, will fully answer for the preservation and culture of a sugar estate in the island of Jamaica in particular, and in the West India islands in general ; maintain it in vigorous prosperity to yield gainful returns, and preserve the capital of the proprietor.

The first thing, I presume, to be considered is, in case of the proprietor's nonresidence in the island, how to fill up the place of a resident attorney with a person of skill, industry, perseverance, and in-

tegrity, in discharging not only the duties
of a planter, in directing and planning,
but to keep fair and equitable accounts of
the various transactions of the estate, with
respect to its culture and incidental transactions, in as simple a manner as the
nature of things will admit of. There
should be laid before the proprietor, in
plain legible terms, the accounts of the
estate to the end of every year; a list of
the slaves and stock, with their increase
and decrease; the cultivation of the estate,
with the returns from plant and ratoon,
in curing house hogsheads of sugar; the
number of acres in cultivation of canes;
returns of rum; the condition of cane,
grass pieces, and provision grounds; the
quantity of acres laid down in a table, whether plant or ratoon; their condition, and
when fit to be cut; the names and number
of the white people resident on the estate,
with their occupations and salaries; the
different island accounts, whether paid or
unpaid, as they are presented; the ship-

ments and appropriation of the crop; what balance of the crop there still remains on the estate, or at the wharf, not yet appropriated or shipped, and the list of clothing and salt provisions served to the slaves; jobbing and tradesmen's accounts, &c. &c. These simple accounts are easily kept, transcribed, brought home, and presented, for the satisfaction of the proprietor, every year : he also keeping such a memento by him as should refresh his memory and satisfy his mind. The person thus commissioned, confided in, and depended upon for punctuality, celerity, and truth, should be one who will undertake the charge as a travelling agent, perfectly assured, under all circumstances, that he would fully answer every intention of his appointment, and endeavour to bring the crops to a speedy market, at one-fifth the expense of a resident island agent, and few mortifying disappointments; far less island contingencies and disbursements. Every thing should be brought within the mental scope

of the proprietor by regular returns, and
duties duly and efficiently performed ; the
entire of his crops brought or sent home
to him, that he may appoint his own broker
to dispose of them ; that he may circulate
his own bills of exchange, and discharge
his island disbursements by their means.
His list of supplies should be brought home
by his travelling agent, who should inspect
on the estate, that only what is really wanted
is sent for, and likewise the nature and
condition of the supplies shipped. He
either should go out in the vessel that
carries them, or follow shortly after it
arrives in the island, in time to detect any
damage or fraud, and without delay have
them brought on the estate. He should
look minutely whether his former orders
have been carried into effect with precision ; examine the condition of the slaves,
stock, and culture ; make his remarks on
any errors which have crept in through
neglect or otherwise, which demand censure or amendment. He should order the

requisite improvements, arrangements, and tillage of lands; have stock purchased, if necessary; old or infirm cattle fattened or sold off; buildings repaired; lumber procured for the entire crop; the slaves served with clothing; the salaries of the resident white people paid by bills of exchange sold for that purpose; the different accounts made up; the produce on hand shipped in good time and order, without loss by ullage, &c. &c. All this can be performed, by a stay of three months in the island, by the travelling agent; and the same person may return to England in the vessel in which the produce is shipped, or be the harbinger of the favourable news in some other vessel. Affairs taking this turn, corruption would have scarcely time or opportunity to creep in; the net proceeds would be in the hands or at the disposal of the proprietor; the overseer and white people resident on the estate would live in a state of anticipation as to the time of visitation. Anxiety for the preservation

of their situations would keep them as-
siduous, sober, and attentive. Having done
their duty, they would be confident of
support, preferment, and reward. Their
example would diffuse a spirit of emulation
in others; their health would be preserved;
their constitutions unimpaired by excess;
they would be comparatively happy in a
foreign tropical climate; grow wealthy by
degrees; and at last, saving a little in-
dependence, would return home, either to
spend the remainder of their days in comfort and plenty, or meet flattering thanks
and rewards from the proprietor for their
fidelity and good conduct, appointing them,
perhaps, to a lucrative situation, as travelling agents; fit to be entrusted, having
undergone the various duties of planters
with patience, temper, experience, and
competent knowledge. Thus they would
combine credit to themselves, and advantage to their employers.

It now comes to be enquired into, what
may be the amount of annual savings

to the proprietor, by employing such a person in the character of a travelling planting agent. The present mode of requiting the resident agent is, principally, by five per centage on all sales, either in the island or on shipments home: that of the travelling agent should be by a salary proportioned to the number of hogsheads of sugar the estate, upon an average of one year with another, is capable of making: from 100*l.* to 150*l.* sterling per annum, with his travelling expences likewise defrayed. The income to the resident island agent would be nearly 500*l.* sterling per annum, besides his occasional residence on the estate, and the establishment of a large house, with its attendant servants, express mule-boys, grooms, grass, corn, and what good things the estate can likewise furnish for his table, &c.; making another additional expence, out of the estate and the proprietor's pocket, of 500*l.* a-year more. As to the mode of payment to the travelling agent, I would presume

to provide for it thus : the fixed salary of
100*l.* or 150*l.* per annum for each estate
should be paid him by the proprietor of
such estate as he may be employed for;
the travelling charges to be paid or defrayed by a number of West India or
Jamaica proprietors, who would confederate
together, and form themselves into an as-
sociation for their mutual interest, to have
the business of their estates transacted and
managed by such travelling agent. By
such a union the travelling expenses would
fall light on each individual, perhaps
never exceeding 35*l.* sterling per annum
for each estate, thereby making the expenses of salary and of travelling not to
amount to 200*l.* sterling per annum for
each estate, and thus rendering the employment worthy the acceptance of a fit person.
The travelling agent having but the short
period of three or four months to stay on
the island annually, and having his time
wholly engrossed between the different
estates, inspecting, executing, ordering,

and diligently providing for the future welfare of the different properties, collecting accounts, &c., no establishment of a large house would be required. The ordinary fare of the overseer's house would be found fully suitable to his expectations. His intentions would be directed to the completion of his business by a stated time ; indulgence, satiety, excess, or voluptuousness, he would neither think of nor care about; his retinue would be small, his appearance unostentatious, his notions unassuming and humble. He would command without arrogance, and dictate with affability and reason. He would be welcomed with joy, respected during his stay, and taken leave of with regret. He would have the good wishes of the white people and the slaves attached to the estates under his management. They would rejoice at his welfare and return back to them. This would give a saving to the proprietor of at least 600*l.* or 700*l.* per annum, without the danger of a sinking fund, or the rapacity

carried on, under the garb of a pompous
resident island agent, with the veil of
treacherous peculation, and screened by
near 5000 miles' distance. The sugar and
rum crop is accounted for fairly, because
shipped home and sold there, under the
eye and superintendence of the proprietor
himself; sound and durable lumber is provided for the entire of the crop, making
an assurance of the produce being shipped
and received in good order, or the captain
and owners of the vessel made account-
able for it, taking care previously to effect
the necessary insurance on the property.
No unnecessary millwright, mason, and
coppersmiths' bills should be allowed;
common-place surveying should be annihilated; extraordinary jobbing abolished,
with its frequent bad incomplete work.
Wharfage accounts should be accurately
ascertained, with divers other incidental
expenses and losses retrenched. No more
supplies should be sent out than what are
absolutely necessary, making another saving

(speaking within bounds) of at least six or
seven hundred a-year to the proprietor.

When the travelling agent arrives in
England, his accounts should be presented
to each proprietor or employer that he is
commissioned by, either at a general board
or individually, in order that they may be
audited by a person appointed for that purpose, either by the board in general or any
individual proprietor. The whole should
be entered in a regular book, a copy of
which should be always kept by each proprietor, for his satisfaction, that he may be
able to refer to it at all times. The travelling agent should keep a book of accounts
corresponding to this. This being done to
general satisfaction, the next occupation of
the travelling agent is to have the different
merchants, tradesmen, &c. advised of the
supplies wanted to be shipped by a stated
time for each estate ; to have them collected,
inspected, marked, shipped, and the bills
of lading signed ; one to be kept in England by the proprietor, one to be trans-

mitted under cover, by letter, to the overseer on the estate, and one to be kept in the
possession of the travelling agent himself:
so that all parties interested therein may
have one to advise and guide themselves
with. It is presumed that this combination
of circumstances, this rule invariably acted
upon, this chain of business linked together
with regularity, and carried into execution
with celerity, assiduity, and vigour, will
meet with the approbation, support, countenance, and adoption of the majority of
the independent, impartial planters.

Having thus treated and descanted upon
the circumstances essentially necessary to
the well being of a sugar estate in general,
and particularly in the island of Jamaica, —
by changing the administration of it from a
resident to a travelling planting agent, with
its consequent preservation and savings,
from tottering ruin to healthy animation,
with its prosperous emanations from nature,
aided by integrity, assiduity, and experience,
— I will proceed now to treat of and expatiate

upon what I presume should be the study
and practice of planting and carrying on
the business and duties of a sugar estate,
in the West India islands, interspersing
here and there remarks upon old customs,
with anecdotes by way of illustration,
which may prove instructive and entertain-
ing; many of them, I dare say, living in
the memory not only of some people in
Jamaica, which they know to be truth,
but some have come to the knowledge of
persons residing in England. I shall indulge in no personalities; my design only
being to do good, all intention of giving
offence is disclaimed: impartiality is aimed
at, without invidious scandal: such motives
must not be attributed to me, as I absolutely disavow them.

As much depends upon the nature, disposition, qualifications, and disposal of the
white people immediately living on the
estate, not only for managing, superintending, and protecting the property, a main
object to be considered is, the selection of

such for the relative duties they are required to perform ; taking care to make
it a primary and most important rule, to be
wholly free from bias or prejudice, in the
employing of men, from any of the three
sister kingdoms, in preference one to the
other. Those of general good character,
for steady, sober, industrious habits, not
given to sudden or violent gusts of passion,
yet of lively, cheerful dispositions, delight-
ing in activity, with corresponding vigour
and health ; ingenious, and susceptible to
improvement, yet not apt to catch at every
trifling experiment, should be preferred.
The overseer should be a man of intelli-
gence, tempered wih experience, naturally
humane, steadfast in well devised pursuits,
not given to keeping indiscriminate company, or suffering his subordinate white
people to do so, thereby vitiating their
manners. His business hours will be fully
occupied by the concerns of the estate, his
leisure ones in the innocent enjoyment of
some domestic amusement. He must be

kind and courteous to the young men
under him, but giving or allowing them no
opportunity to treat him with disrespect;
attentive and hospitable to respectable
strangers; cautious and wary how he suffers strollers to tempt his benevolence. He
must not capriciously or suddenly discharge his white people (as is very often
the case), taking care that no envious or
jealous sentiment or idea arises in his mind,
if his young men have merit on their
side, or are caressed by their superiors.
He must keep the slaves strictly to their
work, yet not imposing on them unusual
hours, or inflicting punishment for every
trifling offence; but when punishment for
crimes is necessary, to temper it with prudent mercy. He must be attentive to their
real wants, not suffering them to teaze him
with their trifling complaints, or tamper
with him by their arts, but promptly satisfy
them, by enquiring into their serious grie-
vances. Above all things, he must not
encourage the spirit of Obea in them

(which is horrible), or dishearten them by cohabiting with their wives, annulling thereby their domestic felicities. He must not suffer their provision grounds to be neglected, trespassed on, or ruined, or their houses to be out of repair or uncomfortable; for it very often happens, that well disposed slaves by such freedoms taken with their wives, their well established grounds ruined by thieves or cattle, their domestic quiet and comfort intruded upon, or their houses rendered uninhabitable by storm or casualty, become runaways. Their conduct influences others, till at last the strength of the estate vanishes, the evil becomes notorious, and the plantation, of course, becomes neglected. The magistrates are then obliged to take this growing evil into serious consideration. Hunting parties are sent out (perhaps with little success) to bring in the fugitives; martial law is at last proclaimed throughout the diseased district; all sorts of people are harassed; public trials are instituted; some of the runaways are never

caught; others who are brought in undergo trial, and are convicted and sentenced
to death or transportation for life. All
these proceedings might be avoided by well
timed, prompt attention at the beginning
of the evil. The overseer loses his character and situation by such misconduct. He
seldom regains them, but often sinks into
contemptuous disrepute; or if he main-
tains them, it is by the stern will of an
all-commanding resident island agent, who
has promoted and sanctioned this odious
and cruel policy. But what is still more
intolerable, still more heart-breaking, and
calls loudly for public justice, is when those
very provision grounds, those very negro
houses, which are nursed and reared by
the painful toil of the negroes, by months
and years of indefatigable labour, those
temples of their present and future happiness, are, by the despotic mandate of a
ruthless resident island agent, contrary to
the island laws, desperately entered into by
a gang to his purpose, the provisions de-

stroyed, the houses with their contents
pulled down, levelled to the ground, and
burnt. It may easily be conjectured what
will be the consequence, when the slave
views his beggared starving family, his
comforts fled, his happiness annihilated
and expiring. An example of this profligate conduct, and its direful effects, was
notoriously evident about the year 1804 or
1805, as practised upon some well-disposed
slaves, belonging to an estate situated in
the mountain, from whence flows part of
the copious stream of the Rio Nuova, over-
looking the Vale of Bagnalls, in the parish
of St. Mary's, not far from the proximity
of that parish to St. Thomas in the Vale.
Such unrestrained free will was known to
have taken place in other parts, and the
following authenticated account has come
to the knowledge of the author.

In the parish of St. Ann's, at no great
distance from the surgy course of the turbulent roaring river, an estate presents
itself to the traveller's view, flanked by

hills, with an aspect to the sea, of no great magnitude, otherwise than for the noise it made, and the tragedy that was acted there; but it remains in the record book, kept of public transactions in that parish, respecting the horrid fate of an overseer who lived there, and the terrible consequences of the conduct I have referred to. I would fain draw a veil over this lamentable event; but by disclosing it, warn others not to tread in such fearful paths, when human nature, however depressed or subjugated, but stung by the anguish of a poison directed to its very existence, will form some desperate design to rid itself of its torment. Alas! so it was in this case: the overseer was reputed to be a man of ferocious manners. By some accounts he had been engaged in the parish of Trelawny to repress the Maroons. The negroes on the estate I now allude to were said to be unruly and turbulent in their manners; and the resident island agent, living at a considerable distance from the estate, thought such a man the best to deal with them. How vain and fruitless are the speculations of some men! Perhaps if he had sent a person of a contrary temper to manage this testy set, they might have been reconciled to their duty, and the disasters that ensued might have never happened. Some mulatto slaves upon the property, as tradesmen, with others, this overseer was in the habit of unceasingly carping at and punishing, for what he thought neglect of duty, upon being threatened, betook themselves to flight, in order to save themselves from his resentment. He then applied himself to the diabolical, ruinous plan of inducing them to return home, by rooting up their provision grounds, and pulling down their houses in despite of the island laws : they still remained out for some time, till the overseer became somewhat sick, and they thought the storm abated. His disorder became less, and their fears returned, as they thought him implacable, — and they conspired to put him to death with secrecy

15

and despatch. He was of considerable
stature, with great strength. In the dead
of the night, when his dwelling house was
divested of its servants except one female,
who either knew and approved of their
design, or they enticed out of the room,
these mulatto men slaves, sssociated with
others who came there. The overseer was
asleep in bed with a young child of his
own, little thinking of, or ever surmising
his approaching fate. The noise of their
entry startled him : he grasped a hanger
that was near his bed ; but before he could
see them, or make an effort with it for his
delivery, they rushed on him in a body.
A violent scuffle ensued for some minutes.
After knocking some of them down, he
often plunged for his escape, but they still
pressing upon him, he at length became
exhausted. They then caught him by the
legs, threw a rope over his head, round his
neck ; and while some dragged him by the
legs, others swung the rope round the bed-
post, and thus, by pulling contrary ways,

they soon strangled him. They then deli-
berately went to his trunk, took out clean
linen, and after washing him and the room
from blood, or any thing that might cause
discovery of the murder, they put on him
the clean linen, and laid him in the bed
again. They then retired, congratulating
one another on thus effecting their design,
as they thought, without discovery, and with
impunity. When he was discovered in the
morning apparently insensible, the neighbouring doctor was called in, who slightly
examined him and pronounced him dead;
and as he had been sick some time, it was
given out and believed he died suddenly.
Some months passed over before this horrid
event was known, which was discovered,
by a quarrel among some of the perpetra-
tors of the act, and the female servant,
who had been in the house when the
murder happened. It came to the knowledge of some neighbours, who soon had
those people arrested. They were tried
and convicted. The female servant, with

one of the conspirators, was found sufficient
evidence against them ; two or three were
publicly executed, their heads cut off, and
stuck upon high poles on the estate, near
the high road. A shocking spectacle, ex-
citing disgust, and branding the estate ever
afterwards with murderous, infamous ap-
pellations.

An overseer should be a man of settled,
sober habits, presenting a gentlemanlike appearance, keeping a regular, well supplied,
comfortable table, without profusion, (which
the internal means of an estate, in small
stock and provisions, generally afford, if
attended to,) not only for himself and the
white people under him, but for the benefit
of such sick and convalescent slaves as require salutary and restoring nourishment.
What with Irish salt provisions, &c. sent
out in the annual supplies, together with
live stock and provisions raised on the
estate, adequate means are usually obtained
for this purpose. The overseer should be
attentive to the white people under him,

E

that their rooms, linen, &c. are regularly kept clean, showing an example of cleanliness in his own person. He should suffer them to sit, after business hours, in his company, instead of morosely banishing them either to their own sleeping-rooms, or to a distant dark part of the house, till meal-time is announced, which induces them to take gross freedoms with the slaves in the house, and meanly assimilates their manners to such company. This, alas! is too often the case. This has sprung from the weak, envious, jealous tempers, too frequently indulged; and has been cherished as an old custom, thinking that by keeping young men in fear, and at an awful distance, it added dignity to the overseer's station. Thus by keeping them ignorant, by heaping contempt upon them, they have swelled and rivetted their own consequence and importance in the minds of the slave population. Where such conduct predominates, the young men are generally unhappy, unsteady, given to drunkenness,

quarrelling, leaving their situations, caballing, &c. The estate suffers by it, the
number of white people is incomplete according to law; the overseer is generally
obliged to accept of the services of vaga-
bonds, because few respectable young men
will live with him, his character being
stamped by such harsh, uncharitable features; and especially as they are liable to
be discharged at every gust of his caprice.
His temper is soured by frequent casualties
of this nature, and vents itself often with
terrible consequences upon the slaves under
him; whereas the contrary would be the effect
of an amiable deportment to the subordinate
white people. Feeling themselves comparatively happy, it would be an incentive
to them to pay every possible attention to
their various duties, to merit the overseer's
favour by giving him satisfaction, to preserve their situations, and acquire a good
character. They would cultivate respect
for him, serve and obey him with alacrity,
intuitively and practically become good

planters, secure the estimation and good opinion of the overseer ; and thus pave the way to their own promotion, and furnish the district, or the island, with deserving, provident managers.

Another difficulty is created by the corrupt, viciously-disposed, criminal views of some overseers, of what they call the old school ; and that is, a determined antipathy to young men sent out to be apprenticed or indented to the estate of the proprietor. To such I have ever found (if they were ever so well inclined) a strong dislike manifested by such men, whose habits of depravity render them ever weak-minded, ever awake to, or dreading representations of their conduct being sent to England, anticipating fatal results not only to the estate, and the management of it, but finally ending in their discharge. This malignant temper principally rests with overseers of the old school, fomented indeed by the precepts of a resident island planting attorney, who likewise is averse to

the introduction of young men, by whom
his conduct is observed. Such young men
are often ill treated by their superiors : their
virtues are of no avail, or are represented
as vices; their vices are magnified beyond
the power of redemption; their lives become scenes of cruel taunt, neglect, and
reproach. This is their portion, till matters
are rendered so insupportable that they are
often obliged to sue for their indentures
to be given up, or run the risk of leaving
the estate previous to the time of their servitude expiring, and seeking employment in
some distant parish. Instances of this kind
I could here easily refer to, and confirm, as
having happened in Clarendon, St. Ann's,
St. Mary's, &c. The bosom of the parish
of Clarendon produced an extraordinary
case of this kind, as likewise that of St.
Ann's. At the descent of its mountains,
not far from the capital of that parish, some
young men were sent out duly apprenticed,
and placed upon the estates, with the watch-
word to the overseers, which was, beware

of these seers; keep them under; allow
them no liberties; bring them up rigorously
to their duties; give them plenty of field-
work in all weathers; coarse diet; place no
confidence in them; give them no rest by
day, or comfortable repose by night; sound
in their ears the slavery of their condition;
make them be so sensible of that condition,
that misery may haunt them. Thus wretch-
edness compelled them to leave the estate.
These young men thus suffered; but it
happened that this compulsory banishment
did not turn out as fortunate for the resident island agent, or his overseer (the
ready executioner of his will), as they
imagined. Something went wrong at court,
something was found erroneous in the
management: fresh powers of attorney
came out; a fresh administration took place.
Those favourite spots which used to be the
residence of the island agent for months,
with its emoluments, '&c. were painfully
longed for again; and after some years of
disturbed patience, hoping by his seeming

disinterested urbanity, and apparent Christian-like conduct, after vilifying these young men in every company he went into, and depicting them in the most atrocious light, as having no abilities as planters, he helped them to overseers' stations: this he did with the hope of once more procuring by the effects of their gratitude, his re-establishment on the property. I must here pay the tribute of praise to numerous young men, who have come out to settle in the island of Jamaica as adventurers, who by their natural good disposition (though unrecommended) have attained to be competent planters, have been excellent overseers, and an ornament to society: many I could name; but this would be considered as invidious by others.

But although sending out such young men under indentures has come rather into disrepute, as they have been represented, by letters to the proprietor at home, as incorrigible, idle, and worthless, yet I will venture to assert, and vindicate the assertion

by convincing reasons, that young men sent
out from home to be indented to estates, are
the fittest for the subordinate duties of them.
And first let me premise, that young men
of moderate education, whether as ap-
prentices or otherwise, are best adapted
to be of effectual service to the estate,
and more likely than illiterate blunderers,
not only to behave themselves well, but to
rise to reputation and distinction as good
planters. It is not necessary that men in
the capacity of planters should be classically
educated, but it is consonant to reason to
have them qualified as English scholars, as
it enlarges their minds, and makes them
capable of their duties. This must, of
course, be a great acquisition to an estate,
where books are kept of annual and daily
occurrences, (a great relief to an overseer,
who sometimes knows very little farther
than the acquirement of signing his name,)
and when time accomplishes their servitude,
or makes them fit to take charge of an
estate, it will make them more esteemed

by their employers, and respected by the community at large. Many things arise in the exercise of an overseer's duty which require intelligent, nay even sometimes refined talents, to go through with. The best persons, I presume, to be sent to Jamaica, or any of the West India islands, who are to undergo the minor duties of an estate, are young men from seventeen to twenty-one years of age, who are not too old to abide by command, constraint, or instruction. The body and the mind are vigorous and pliant at the former age, adapted for toil, less captious, or averse to controul or advice, and capable of performing the duties of a soldier; an effective militia, for the preservation of the island against invasion and insurrection, being kept up of men in every station there from sixteen to sixty years old. The legislature of the island of Jamaica has wisely imposed a deficiency law which enacts, that a specified number of white people, capable of bearing arms, of the before-mentioned ages, shall be resident and employed upon each property, according to the number of slaves and working stock; or be assessed a fine for each deficiency. This militia has had the merit of very often performing the most salutary, effectual, and gallant services to the island, not only as local, but marching troops. Young men of this class, after satisfactorily serving a period of four or five years in the planting line, may be deemed not too young to be given in charge of an estate. Young men sent out after they pass the age of twenty-one years, are apt to entertain notions of insubordination; the burthen of servitude sits uneasy upon them; liberty, the darling object, the noble pursuit of mankind, is often uppermost in their minds. Not that they are less virtuous for harbouring the sentiment and passion, but that they are less subject to controul. The commanding, and sometimes the arbitrary orders of the overseer are often slighted by them. They are prone to be disrespectful, guilty of retort, only carelessly or half

15

obeying his lawful injunctions. This is a crime ever to be avoided in the planting business, as carrying with it a want of deference, and setting an example of insubordination to the rest of the white people, and the knavish, artful slave. This gives rise to unpleasant disputes; the overseer complaining of the presumption and neglect of the young man, and he urging in vindication, the harsh, inflexible conduct of the overseer, judging by his too liberal notions, that he has been hardly dealt with. He then grows sullen, and often impertinent, conniving at, and provoking aspersions on the overseer, who generally has some friend or enemy to inform him of it. The breach is widened, and at last, to wind up the affair, the young man is either removed to some other estate in the same employ, or voluntarily withdraws himself, to avoid this unpleasant termination of the dispute. It would be better, therefore, to have young men sent out of the age just now specified. And having now, I hope, described, with

conviction to my readers, what kind of
white people are best adapted for the profession of a planter, I shall take a transient
view, with a few cursory remarks, of some
other white people, found often to be
necessary for a sugar estate; that is, tradesmen, who either come out indented to, or
hired upon the property. Young men are
frequently sent out indented to an estate
to conduct a tradesman's department; such
as millwrights, carpenters, coopers, coppersmiths, masons, &c. They have generally
served their time, or the principal part of
it, to such particular trades in the mother
country, before their arrival in the West
India islands. They are seldom under
twenty-one, and very often forty years of
age when they arrive there. They are influenced to go out under this character,
either by the temptation of a liberal salary,
to execute some difficult design, or have
failed in getting employment at home;
some are men of sound judgment, but a
great many more mere adventurers. Their

employment being rather foreign to an
overseer, they for a length of time baffle
his power, and oppose his disposition.
Sometimes the work is executed in a master-
ly manner, though great delays attend its
completion ; but it is often done in so bad a
style, as not to answer the end intended, after
a great lapse of time, waste of stuff, the
labour of a powerful gang of negro tradesmen (the flower of the slave population)
lost or made of little avail, and the cattle
half destroyed by dragging timber, and
heavy materials. Frivolous excuses, with
contingent insuperable obstacles, are assigned as the cause of this failure, waste of
time, labour and materials. They industriously, but falsely, strive to blame the
overseer for the mischance, after harassing
his mind for months with some want or
other, expending labour and materials un-
profitably, reducing the cattle, delaying the
crops, and at last oblige him to call in some
competent tradesman to render things complete. Dreadful is the situation of an estate

so circumstanced. This stipendiary ignorant mechanic, in the mother country perhaps acquired the habit of drinking, and
follows the practice with avidity in his
new situation. It would be much better
then, I presume, casually to employ, when
wanted, the tradesmen of eminence throughout the Islands, to execute any arduous or
difficult jobs, provided their charges were
moderate, than to be burthened with the
maintenance, and ill-earned salary of such
useless agents. The negro mechanics
attached to the property, are generally
found sufficient for common-place work or
undertakings ; they are even known to be
capital workmen sometimes, and have often
great knowledge and skill.

I must here beg leave to observe, that the
valuable island of Jamaica produces in great
abundance, a variety of the choicest hard
wood timbers. Millwrights, and carpenters,
sent out from home, are entirely ignorant of their nature, names and texture when
they arrive in Jamaica ; for the knowledge

of these they must be indebted to the experienced slave. If they were sent into a wood, to make choice of such as were necessary for durable work, they would as likely cut down a huge cotton tree, a birch, or any other soft wood, as a cogwood or bullet tree. In the same manner would a mason, sent out from home, if told to go to the woods, and have a lime-kiln made, or built, be totally at a loss how to begin or construct such a work. Moreover, they are incapable for a length of time, of understanding the dispositions of the negro tradesmen they are put over; they either make to free with, or are violently tyrannical over them, which unavoidably causes default in the slave, and tiresome (perhaps punishable) complaints to the overseer.

Some proprietors (probably more through instigation, than the natural bent of their own dispositions) are induced to send out indented ploughmen from home, to use that implement in turning up the land for tillage and the culture of canes. I shall treat more at large in a subsequent place, as to the utility of the plough in Jamaica. But my present object is to point out, that such persons as are usually sent from home, seldom answer the purpose intended ; they generally turn out as great an evil as the indented tradesman to an estate. They are mostly very illiterate, ill-informed men. They have been in the habit of driving the flexible, alert horse in the plough, with proper gear to work with ; whereas in Jamaica, the dull, heavy, stubborn ox, or steer, is the animal they have to train up, work and contend with, coupled with tough, heavy yokes and bows, and seldom used to any other work than waggon, cart, or mill-work. Besides the negroes are often as ignorant of this mode of work as the ox himself; so that the white ploughman, what between the disposition of the negro and the ox, finds himself dreadfully harassed, every thing nearly depending on his own exertions, in a large field of ten or fifteen

acres to turn up, so that the overseer may
have it planted in good time, and that the
plants may not spoil through age, and be
unfit to put in the ground. What is the
result of this experiment? the white
ploughman is considered as incompetent:
often through fatigue of mind, or body, he
becomes sick and feverish; the cattle are
half killed, the ground only half ploughed,
and the overseer obliged to draw off from
other important work the most effective
field-gang of the estate, and resort to the
alternative of finishing the turning up of
the land, by digging it with the hoe; the
mode mostly used and best understood in
Jamaica. Thus generally ends the career of
ploughmen sent out from home.

To sum up this dissertation upon the
kind of white people, competent in body
and mind, to perform and go through the
duties of a sugar estate in the island of
Jamaica, the author, instructed by nature
and experience, inclines to close it, by
a very few desultory remarks, springing

F

from the situation in which white people
are placed as planters. Few stations require more real comforts than should fall
to the lot of the white man, exercising the
profession of a planter in that island.
Emigrated from his native country; alien-
ated from its joys, and the seasons which
bring the gifts of bounteous Providence to
his gladdened heart, he lands in this un-
known climate, invigorated with health,
a hale constitution, with little thought of
the fatigues, disappointments, and dangers
he is to encounter; he is placed instinctively
as it were upon this land of delights, this
never-fading verdant soil, ignorant of its
products, the variety of its population, its
allurements to pleasure, its resources to
sustain nature, its customs, blessings, and
inherent, but lurking, fatal distempers;
surrounded by strangers, he scarcely knows
what to do, or how to act; it should then
materially behove the people that he comes
to reside among, who are experienced in
the affairs of the island, to be solicitous
about his welfare, (in fact about their

own general good,) and by degrees to
bring him on to bear the toils of his situation, to give him wholesome advice, to
act in a friendly manner, not by indulging
him in his appetites or propensities to irregular, inordinate practices, but that seeing
his necessities are kindly and wholesomely
attended to, he should not be entrusted
with the keys of the different stores, till
he has sufficient experience to keep them
with safety, and be aware of the subtle
craft of the negroes. He should be accustomed to superintend for some time, as
part of his duty, the rearing and feeding
of small stock, (a most essential point to
be attended to on every property,) the
keeping accounts of them, counting and
seeing the cattle and mules dressed of
their sores and wounds. I must here observe, that many overseers are so profuse
at their table, as soon to wear out, and expend, the entire of the small stock, leaving
themselves, and the white people living
with them, nothing but the salt beef and

pork in barrels to live on, nay, even sometimes nothing but salt herrings.   It is then
that the wretched situation of such persons is to be deplored; their very appearance denotes the beggary of their way of
living.   Squalid, and often ulcerated, they
are tempted to exchange salt provisions,
with the slaves, for a meagre fowl or
duck, to procure a salutary meal.   When
once the breed of small stock is too much
diminished or destroyed, it requires great
perseverance, patience, expenditure of corn
and time, to restore a sufficient number,
to give the means of comfortable subsistence.   But if this is properly established,
with care and economy, they are a lasting
source of plenteous nourishment to the
healthy, (whom they preserve,) and to the
sick and convalescent in the time of need.

I shall conclude my observations on
white people, and their treatment, with re-
marking, how necessary it is for every
white man on an estate or property to be
possessed of a horse in his own right, in

order to enable him, not only to discharge his public duty with ease, but to give him occasional recreation, with accustomed respect. To attain this object, a great part of a book-keeper's or a subordinate white man's salary is laid out in the purchase of one, which he must of course be most anxious to have taken care of, as the loss of it reduces him to the lamentable condition of comparative beggary and destitution. Capricious fate may decree him to be discharged or discarded from the estate, for some trifling or heinous offence. At this critical time, he is without this necessary companion to bear him through his difficulties, and often compelled to leave the estate on foot. Then, alas! calumny marks him as a vagabond, unfit to be employed. He is slighted and treated with little pity by the community. Instead then of treating the horse of a book-keeper with indifference or neglect, neither allowing him good stabling or feeding, the overseer should humanely direct his attention to the wellbeing of his book-keeper's horse, recollecting, that he was once in that situation himself. How pitiable then must be his condition, if he loses this valu- able animal, his very respectability depend- ing on his being possessed of one! This is a cardinal point to be attended to by an overseer, in his treatment of white people, for by paying proper attention to their comfort, he makes an indelible impression on their mind; his head will recline on his pillow with consoling ease, and it will be a lasting, beneficent trait in his character.

We ought to consider the perils of a white man living as a planter in Jamaica, bringing with him a robust, healthy, European con- stitution, the hardships he has to undergo (which quickly undermine the constitution) from being out in the field in all weathers, in such a climate. In crop time, one half of his nights are spent in a sultry boiling- house, at a distance from his dwelling, and he is called out of his warm bed in the middle of the night, exposed, perhaps, to nstant wet or storm, or half-way up to the knees in mire in his passage to the works. The fevers which ensue oblige him to pay, out of his own pocket, for medical advice, which comes very high there. His loss by the death of a horse, nothing but time and hard labour will repair, and put him in possession of another. He has to fit himself out with regimentals, and find himself with decent clothes; and every article of wearing apparel being nearly three times as dear there as in the mother-country, and the consumption twice as much, adds considerably to his burdens. I presume to say, if the considerate, humane-minded employer reflects on these things, he will be inclined rather to add to than decrease the salaries of the resident white people, on a plantation or sugar estate. Besides increasing their salaries according to their merit, good conduct must be encouraged. The business of the estate will be better and more effectually conducted by men of good than bad character. In the long run, the happy effects

will be productively apparent, savings will
accrue, satisfaction and happiness will be
the issue. It is much easier to get rid of
bad young men than to procure good ones;
so that when there are well-disposed, intelligent young men on a property, it would
be well to keep them there, and show them
the advantage of being steady, by reward-
ing them for their services, with a small
increase of salary. I beg leave to assert, that no overseer on a sugar estate
should have less than 200*l.* currency per
annum, (whatever more he may obtain,)
and no subordinate young man, in the
character of book-keeper, less than 80*l.*
currency per annum, and whatever more
he may receive according to his merit. The
addition of 10*l.* or 20*l.* per annum to each
would scarcely be felt by the proprietor,
and the property benefited some hundreds
of pounds a year by it.

It is likewise a source of great additional
comfort for every property to have a well
kept, plenteously stocked kitchen-garden

on it, which can always be established and continued at a trifling expense, an invalid negroe generally being appointed to such work, with a comfortable cabin to live in, to preserve seeds and plants, and guard it from thievish invasion. It is almost needless to say, that even in a temperate climate, the use of garden-vegetables, and pot-herbs, is considered as most essential requisites to health and enjoyment; how much more so must such salutary productions be to men living in a land exposed to the effects of a parching, tropical sun, whose blood is in a state of almost constant fermentation, whose exhaustion is excessive. If their food is not qualified with the purifying, grateful influence of such ingredients, the system becomes morbid, extreme languor ensues: the corrupt, latent seeds of disease burst forth at length into some terrible distemper or malady, proving fatal to many. Add to this, the practical economy of a table furnished with viands of this kind. The islands in gene-

ral afford abundance of natural productions of this sort, the roots and seeds of
which are easily procured, which, if sown
or planted properly, nursed in their infancy,
and kept even moderately clean, will be a
perpetual source of wholesome nutritious
supply. The Indian kale, ochro, quash,
peppers, ackys, and a variety of pulse,
being natural to the climate, together with
a few fresh European garden-seeds, sent
out regularly every year in the supplies,
which thrive very well here, give an abundance and variety, which few soils or
climates can boast of. This garden should
never be at a distance from the overseer's
house, as his eye and talents are its safeguard, nurse, and support.

CHAP. II.

THE TREATMENT OF SLAVES.

On entering and expatiating on the treatment I presume most proper for slaves,
especially those under the command and
management of overseers in the planting
line, (which is a great and even cardinal
point in the spirit of plantership,) I shall
first advert to the ameliorating laws enacted
by the assembly of Jamaica, which are
recognisable by every owner of slaves,
overseers, &c. living in that island. Than
these nothing more belies the slanderous
representations of some overrating philan-
thropic people, who assert, that their condition, comforts, rights, and protection
from severities, are not attended to. Having
the knowledge and experience of those ex-
isting laws, not only made and promul-
gated, but strictly acted upon, under heavy
penalties of fine and imprisonment, the

violation of them cannot escape a numerous magistracy, or the contumely of a watchful community, I shall venture to avow in the face of the world, that there are no class of people in their sphere, in the universe, whose faults and natural tendency to crime are more abridged and looked over, their wants supplied, their comforts attended to, nay their very existence, when tender infancy, or decrepid old age requires care and succour, more humanely or rigidly looked after. A quarterly return is made to each parochial vestry, establishing thereby an inquisitorial power into the increase and decrease of the slave-population for each property, by which it is seen, in general, whether there is a growing increase of the population, putting a quietus to such groundless fabrications.

I shall go on (without alluding to former custom) into the train of practice I suppose best suited to treat the adult, strong, healthy slave, the youth, the infant, the invalid, and superannuated, classing them according to their different occupations. We

have not for some years imported, neither
is it ever likely to take place, that we should
ave a fresh supply of slaves thus brought
into the British colonies. The old Africans
are daily wearing out and dropping into
the grave; our care is to support the
present stock, encourage healthy propaga-
tion, lessen their propensity to vice, caba-
listic or obea arts, induce them to receive
Christianity, not to excite their hatred or
jealousy by lewdness or wicked practices
with their wives, — a baneful custom ; to
take care that they are regularly supplied
with salt provisions, (which they prefer to
fresh, being good, savoury cooks in their
own way;) comfortable clothing ; that their
houses are kept tenable; their time and
hours to cultivate their grounds not in-
fringed upon; those grounds kept free
from trespass of cattle or otherwise ; that
they be not punished for every trifling
fault, or unmercifully, at any time ; when
really sick, that they be taken into the hospital, under the care of the attending doc-

tor, with proper medicine, nourishment, &c., for them; that their infant children are provided with proper nurses when weaned, kept clean, free from insects called chegoes; a wholesome mess of stewed provisions, with a proportion of garden-stuff, made savoury by a little salt meat, &c., served to these children every day, in the overseer's presence; the invalids and superannuated treated with sympathy; and their sufferings, brought on by either age or infirmity, relieved. By such usage as this, the slave becomes attached to the property he belongs to. He only nominally is such in his own thoughts; his master's property is his freehold; the property cannot thrive without him, or he exist without the property; he gets old in its service; has children to comfort, support, and soothe him when past his labour, who generally glory in their regard to their parents. This is a most respectable spectacle either on the estate, the public road, or at the provision-market. He sinks quietly into the grave at a full old age, and leaves with studious impartiality his little property (of whatever kind it may be) among his children, whom he has trained up to pursue his manner and mode of life. Thus, with the blessing of Providence, insuring to the proprietor a succession of healthy, well-disposed, effective slaves. Casting a look over the European map, I can discern districts, I may say entire countries, styling themselves civilised, which are now ordaining laws for the balance of power, securing their dominions by the specious appellation of religious, Christian-like epithets, when nine-tenths of the population of those countries only have nominal freedom; few of the comforts, protections, and enjoyments of the slaves in Jamaica, and the West India islands, and are in fact the veriest slaves in the world.

The most important personage in the slave-population of an estate is the head driver. He is seen carrying with him the emblems of his rank and dignity, a polished staff or wand, with prongy crooks on it to lean on, and a short-handled, tangible whip; his office combining within itself a power, derived principally from the overseer, of directing all conditions of slaves, relative to the precise work he wishes each gang or mechanic to undergo or execute. The great gang is comprised of the most powerful field-negroes, and is always under his charge. These are the strength with which principally to carry into effect the main work in the field, and manufacture the sugar and rum. There are so many points to turn to, so many occasions for his skill, vigilance, steadiness, and trust-worthiness, that the selection of such a man, fit for such a place, requires circumspection, and an intimate knowledge of his talents and capacity. A bad or indifferent head driver sets almost every thing at variance; injures the negroes, and the culture of the land. He is like a cruel blast that pervades every thing, and spares nothing; but when he is well-disposed, intelligent, clever, and active,

he is the life and soul of an estate. He
very often is an elderly or middle-aged
negro, who has long been so employed.
If it should be so ordered, that a new
head driver is requisite to be put in commission, I must beg leave to lay before my
readers, my opinion of the proper choice
of one. I may err, but hope not irre-
trievably. He should, in my judgment, be
an athletic man; sound and hardy in constitution; of well-earned and reputed good
character; of an age, and, if possible, an
appearance to carry respect; perhaps about
thirty-five years old; clean in his person
and apparel; if possible a native or Creole
of the island, long used to field work, and
marked for his sobriety, readiness, and
putting his work well out of his hands.
His civility should be predominant, his
patience apparent, his mode of inflicting
punishment mild. He should be respectful
to white people; suffering no freedoms from
those under him, by conversation or trifling
puerile conduct. It is rare, indeed, to find

this mass of perfection in a negro; but
you may obtain a combination of most of
those virtues; and as to petty vices, always
inherent in some measure in human nature,
they must be looked over, when not too
full of evil. The junior drivers likewise,
if possible, should be men of this description; but having a good master over
them in the head driver, they will be induced to behave themselves tolerably. It
gives a great deal of vexation to an overseer when he changes his head driver.
Caprice should never have any hand in
such a transaction. The overseer who
thus trifles, who thus stakes the fortunes
of an estate upon mere frivolities, deserves
never to be employed again. The burden
of the ensuing mischief that may happen to
the property, should rest on his shoulders.
Yet it indispensably behoves an overseer
to get rid of, or dismiss a bad head driver;
for such a one he will soon find out. When
ill disposed he will perceive the negroes
likewise so; the work will not be carried on

agreeable to his dictates; things suffer in
general; the slaves run away, or are inclined
to be turbulent; he and they cabal; bad
sugar is made; and perhaps the horrid and
abominable practice of Obea is carried
on, dismembering and disabling one another; even aiming at the existence of the
white people. The root, then, of this evil
must be struck at, and the head driver and
his abettors sent to public punishment.

Another most material person on an estate
is the head cattle and mule man. These
are people of great responsibility, having
under their charge a great portion of the
proprietor's capital, much depending on
them for their safety: bringing the canes
from the field to the mill, for its constant
supply in crop-time, and carrying the crop
to the barquadier. They have to keep the
cattle and mules in good order, and likewise make them perform their duty well.
They should likewise, with the head driver,
have the good of the estate at heart; have
a proper choice of what cattle are best

adapted for field, mill, or road work ; know the temper and abilities of the stock ; the fit and regular time to spell or relieve them with others ; have always a sufficient quantity of proper tackling for them ; the best mode of feeding, and dressing them for occasional bruises, sores, and wounds ; resting those that are lame, meagre, or that are intended for some stress of work : they should be sober, steady, hale, respectable men. Their employment both in and out of crop, should be the working, taking care, and feeding of their cattle and mules. They must not be drafted to other work, putting the cattle and mules thereby out of the pale and exercise of their responsible avocations: an old, but bad practice. Theft is often practised by cattle or wain men, in carrying the sugar and rum to the wharf; likewise plundering the supplies (especially salt provisions) in bringing them up to the estates. Care should be taken, if possible, to have the head cattle or wain men vigilant and honest; that the head mule-man is like-

wise so ; neither taking himself, nor allowing others to take mules off from the
property, for his or their private use or
work, without the overseer's licence. This
is often done, to the great injury, and perhaps loss, of the mules. Thousands of cattle
and mules are yearly destroyed throughout
the island, by the careless villany or connivance of vicious and profligate cattle and
mule men.

The head boiler or manufacturer of sugar
is another slave, whose trust and employment, during crop-time, is of the most responsible kind. He should always be a
person who has an intimate knowledge of
such a process ; the way the cane has been
raised and treated ; the kind of soil it grows
upon ; if that soil has been high or low
manured ; the age of the cane ; the species
it is of ; whether it has been topped short
or long in the cutting ; if it has been
arrowed, bored, or rat-eaten ; giving him
by this perspicuous view, a thorough knowledge of the lime tempering the cane-juice

requires; the time it may take to concoct, inspissate, and be fit to skip into the coolers. He must be impartial in his mode and time of potting the sugar from the cooler into the hogshead, so that it stands the hogshead well, cures properly, lets off the spumy, spurious molasses, without embodying it in the sugar, thereby giving it an open, free grain. He should be an economist in boiling the sugar, without being a miser to the distilling house. He must be honest, sober, industrious, and keep the junior boilers to their work. Such are the qualities, I presume, requisite for a head boiler on a sugar estate. The fairest fruits of a cane field have been destroyed, perverted, and rendered a mass of thick, slimy, dark, sour, cloddy, unprofitable, unmarketable substance, (disappointing the expectations of the overseer,) by an improper choice of such a member, or having a villain for conducting such a business. The labour of negroes and stock have often been lost by this means; the trash-house consumed

or emptied, shipments disappointed, and
the adulterated juices sent to the distilling-
house, where it will scarcely pay for its
boiling.

Other head men, such as carpenters,
coopers, masons, coppersmiths, and watchmen, are next in succession as principal
slaves on an estate. They generally arrive
at their headship, from being distinguished
either by the proprietor, overseer, or some
superintending mechanic, as good workmen. They are found of infinite service
in the various jobs frequently requisite to
be done : for the building, improving, and
repairing of the manufacturing houses, &c.
saving the proprietor (if ingenious, industrious, and sober,) a considerable sum of
money annually, by not having occasion to
call in the aid of an eminent tradesman to
execute the business. They should always
have plenty of materials to keep them
employed; seasoned wood to work; and
the masons and coppersmiths repairing out
of crop any damage done in it.

A head watchman is always a necessary slave officer on every property; but if such a person is not narrowly looked after, or of extraordinary good character, he spends the greater part of his time in gadding about; working a distant ground of his own; harbouring runaway slaves, whom he cheaply hires to perform some work for him; or perhaps takes an effective mule off the estate, to carry some provisions with despatch to market. This is a bad example to the slave population, who are ever prone to catch infection of this kind. To prevent its noxious influence, this man should frequently, in the course of the day, be with, or at the overseer's house; early in the morning he should go round to the watchmen stationed at the works, and see that every thing is as it should be. He should make a report of the state of the business to the overseer; go round to the cane piece, watchmen, and cattle-pens, and observe if any trespass has been committed, or fences broke. At breakfast-time he

should bring the different cane piece watchmen, with their weapons of defence on one
side, and their rat-springs on the other, to
the overseer's house, to see their success
in destroying those hurtful creatures. He
should always have along with him a
number of active smart dogs, trained up
to hunting those animals, who are in immense numbers throughout the cane pieces,
provision-grounds, and ruins; likewise to
chase and catch the freebooting hogs that
are let loose from the negroes' pigsties.
This head watchman should go over the
lines of the property once or twice a week,
through the woods, and strictly observe
that no damage is done there, or loss sustained by trespass; and report the same to
the overseer. He should be ever watchful
that no mischief is done, or trespass committed on the negroes' provision grounds,
keeping the watchmen there most particularly to their duty; and he should take
care that the fences are repaired where
broken, by those who are appointed so to

do. He should attend at night when the head driver waits to get his orders from the overseer, to know the names of the nightly watchmen to be stationed at the works; and, before he retires to his supper, observe if those watchmen are at their posts. Regularly every week, on Saturday or Monday morning, he should have the handicraft watchmen bring an ample supply of well made mule-pads; fine hackled plantain trash, from off the stem of the plantain tree; ropes for mules, waggons, and cattle tackling; trash and dung baskets, lining pegs, rat-springs, &c. brought home, and deposited in the appropriate store, to be had when wanted. By such a rotation of duty, this officer can be extremely useful, and his time well spent. As an incentive to the principal headmen of an estate to do their duty well, or reward their exertions, to those that are most exposed to toil, inclement weather, loss of time by superintending others, a weekly allowance of a quart or

two of good rum, some sugar, and now
and then a dinner from the overseer's table,
will be found of salutary effect. If deemed
necessary to punish for bad conduct or
neglect of duty, such benefits can occasionally be withdrawn.

I now come to call the attention of my
readers to another class of slaves, whose
lot of occupation comes more immediately
under the proprietors' or overseers' eye;
they rank in the capacity of domestics and
house people. I shall first advert to the
hothouse or hospital doctor or doctress, (as
they are termed in Jamaica,) midwives, &c.,
a most fearful fraternity, who in the course
of the year, may do a great deal of good,
or promote and establish an infinite number
of disorders; having, perhaps, in that time,
the whole population of the estate, — white
people, mixed, coloured, and black — under
their care. Acquainted with medicine only
in a superficial manner, if ever so experienced, they never should have the charge
of the bulk of what medicines may be on the

estate; and what they are put in possession of, should be of a simple nature. Of deleterious drugs they should never have the mixing up; and the utmost caution should be observed when they are allowed to administer any such. A few doses of glauber salts, sulphur, rhubarb, castor-oil, camphorated spirits, bitters and plaisters to dress sores and make blisters of, with two or three lancets, a pair of scissars, and spatula, is all they should have under their immediate care. In fact, an experienced, attentive overseer or book-keeper (as is usually the case) will perform cures in ordinary, simple cases, compound and administer the medicine to the afflicted and sick, with little necessity to call in the aid of the practising white doctor, except when danger threatens. Indeed, some gentlemen of that character in Jamaica, are very little entitled to that appellation. They have large incomes from each estate, without doing any good whatever in a year's visiting. However I shall leave

15

such people to the censure of their own profession. An hospital of sick diseased slaves, is a source of great unhappiness to an overseer. Humanity should call forth his attention ; his duty and the interest of the estate should bind him to it, if the strength of the estate is in the hospital, in a manner lying dormant. If a valuable slave remains there lingering, his mind is sure to be tainted ; his work is delayed, and danger may accrue to the capital under his charge. His character may suffer, and his situation become precarious. Every thing conspires in the mind of a sedulous, humane overseer, to do all in his power to free the hospital of its patients, and restore them to renovated vigour and health. A book of medical treatment, especially of such diseases as are incidental to tropical climates, and is applicable to the cure of negro distempers, should be always kept on every property in the West Indies. I would prefer a male to a female in attending the hospital ; and there should always

be a room there, for the comfortable ac-
commodation of such a person, in case his
nightly attendance is requisite. The male
and female patients should be kept separate,
in comfortable warm apartments, blankets,
&c., and a fire place to each apartment.
Nourishment should always be afforded
from the general store, and overseers' table
to those who require it. It requires a nice
discernment and discrimination to know,
who are to be admitted for medicine or
otherwise, into the hospital. The slaves in
Jamaica, ever given to a most flagrant
abuse of whatever may be established, or
presumed to be for their benefit, the whole
population of an estate, (with a few excep-
tions), would present themselves for admit-
tance there, if the house was large enough
to contain them, or their artifices not well
understood, whether they wanted or not
the aid of medicine. Even the wary experienced overseer they will strive to over-
reach and deceive; nay, they will force
nature from its due course, and by a tem-

porary contraction or revulsion of some vessel in their frame, effect their lazy purpose of sitting down in the hospital. Sores they will irritate and keep alive, fresh ones inflict, and medicines swallow with avidity to avoid work, get in there, and enjoy their supine, idle propensities. They always practise upon a new overseer such tricks. Upon such application, I would give them a hearty dose of some simple medicine, and have them strictly confined to the hospital. If no practising doctor is employed for the property, let them remain there for two or three days, and if nothing apparently ails them, send them to their work. If a doctor is employed, let him examine into their respective cases; if not found unwell, send them to their work again; and let a regular hothouse book be kept of what medicines are ordered and administered, when they are taken in, and when turned out.

Midwives are generally elderly women on a property, who attend the breeding-women, in time of child-birth. They are

in general egregiously ignorant, yet most
obstinately addicted to their own way; but
still if they find danger fast approaching,
most probably brought on by their own
tampering, they will cunningly run to the
overseer, tell him of the dangerous case,
and that he should send for the doctor;
and when he arrives, when sinking nature
is nearly lost in the dissolution of the
mother or child, or one or other expires
shortly after his arrival, they dexterously
assert, that if he had followed their advice,
all would have been well. They impress,
by the nature of their office and by such
assertions, such an awe and reverence for
them on the minds of all classes of slaves,
that few practising doctors wish to encounter them, or be called in to assist at a birth,
or give relief to a female slave in travail,
which those harpies attend. The overseer
can do little or nothing, in those trying
cases, farther than afford medicine, restoratives, and nourishment, for which he is
called upon abundantly by these practition-

ers, and which he gives with the freedom of
a father. Encouraging the midwife in her
attention, for the welfare of mother and
child, he gladly has them taken care of, if
living; and consoles himself that no blame
can be attached to him for any failure. If
a happy issue is effected, he rewards the
midwife and mother, and rejoicingly adds
another name, to the list of slave-population in the plantation book.

The house people should always be composed of the people of colour belonging to
the property, or cleanly, well-affected slaves
to white people, who understand the way of
keeping a house clean in that country, the
care of house-linen, needle-work in general,
and cookery. They should be neat in their
persons, without disease, not inclined to
quarrelling or much talking, civil in their
manners, not addicted to steal away to the
negro-houses, neglect their work, to pilfer-
ing or drunkenness. Having such people as
these in a dwelling house, the white people
and themselves feel, that they are com-

H

paratively happy. If an overseer upon every frivolous occasion, (which often happens,) changes his domestics, he seldom is comfortable for an hour; every thing is at variance; a dirty house, tattered linen, waste of every thing, tumult and punishment going on, caballing, conspiracy, perfidy, and attempts perhaps against his life, many instances of which could be related here. When an orderly set is once in a house, they are with little trouble kept to their duty. As jealousy is apt to creep in among the females, the overseer should give them little or no cause for it; it is a raging, unforgiving, relentless pestilence. Whatever needle-work is requisite, such as making and repairing house-linen, shirts, and stockings, making the clothes for the slaves, who have no wives, or are ignorant how to make it themselves; put a presiding house-woman of good conduct over them, to instruct and superintend them; put no temptations in their way, by entrusting them with the store-keys. Give them a small, but not a profuse part of what meals you partake of. Let them have due time, by relieving one another, in the course of the week, to work their provision grounds, and mind their little poultry and pigs, not suffering them to raise them about the dwelling or overseer's house.

The great gang. — Nothing animates the planting system more than the wellbeing of this admirable effective force, composed of the flower of all the field battalions, drafted and recruited from all the other gangs, as they come of an age to endure severe labour. They are drilled to become veterans in the most arduous field undertakings, furnish drivers, cattlemen, mulemen, boilers, and distillers. They are the very essence of an estate, its support in all weathers and necessities; the proprietor's glory, the overseer's favourite, directed by him. Brigaded by its chief field-officer, the head driver, they inspire confidence, and command respect. This gang, composed of a mixture of able men and women,

sometimes amounting to an hundred, should
always be put to the field work, which
requires strength and skill in the execution;
such as making lime-kilns, digging cane-
holes, making roads through the estate,
trenching, building stone walls, planting
canes and provisions, trashing heavy canes,
cutting and tying canes and tops in crop
time, cutting copper-wood, feeding the
mill, carrying green trash from the mill to
the trash-house, and repairing the public
roads, when allotments are to be worked
out.   They should always be provided
with good hoes, bills, a knife, and axes, to
those men who know how to make use of
them.   They should have these tools kept
in the most serviceable order.   They
should be made to work in a parallel line
as they are set in.   The head-driver, his
assistant-driver, and bookkeeper, should
visit each row, and see that they do
their work well.   An animating inoffensive
song, struck up by one of them, should be
encouraged and chorussed while at work;

for they are thought good composers in their own way. No punishment should be inflicted, but what is absolutely necessary, and that with mercy. In bad weather, a glass of good rum should be given to each ; and when making lime-kilns, roads, and digging cane-holes, a small proportion of rum and sugar likewise to each. Their cook should be regular with their breakfast by nine o'clock in the morning, and their salt provisions constantly served to them. Keeping them in heart, they will work accordingly. They should not work them out of crop, either before day or after dark, (a custom formerly practised,) for they are chilly in their nature, and liable to frequent colds, which bring on fevers and pleurisies. A few hours of such work might give a patient to the hospital for a month. It is, when the all-quickening sun has influenced the creation, that the field-negro feels alive to his work, and announces it, by his cheering song, and redoubled efforts. In heavy rain, all orders of field-negroes should be called in by sound of bell or conch-shell. Attention to these remaks, I presume to think, will add to the stability of an overseer's birth, and be a rule to guide him by.

Second gang.—This gang should be composed of people, who are thought to be of rather weakly habits, mothers of sucking children, youths drafted from the children's gang, from twelve to eighteen years of age, and elderly people that are sufficiently strong for field-work. They should have a competent driver to follow and direct them. Their strength and abilities should be ascertained and assimilated to field-work of the second order, such as cleaning and banking young canes, turning trash on ratoon pieces, threshing light canes, chopping and heaping manure, planting corn, cleaning grass pieces, carrying dry trash in crop time to the stokeholes, and such work, requiring no great strength. The mothers of sucking children should be provided with nurses to take care of the infants, while they are at work in the fields, and a hut made in a convenient place, to retire to, in case of stress of weather. One mother out of every four in the field should be allowed to go and suckle her child for a quarter of an hour, then succeeded by others, and so on, that the infants should not want, and those mothers should not be obliged to turn out to work before sunrise, or be detained to work after sunset. They should have a weekly allowance, of a pint of flour or meal, with a proportion of sugar for each child. The mothers and infants should be kept clean, and free from chegoes. A yard or two of flannel and check, should be given to each infant, for a frock and cover, besides their usual allowance of clothing. In all other respects this gang should be treated as the other slaves on the property are.

The Third or Weeding Gang. — This corps, forming the rising generation, from which, in progress of time, all the vacancies occurring in the different branches of slave population are filled up, comes next to be considered. Their merits are great in their sphere. The expectations formed of them are still greater, when contemplated in a future point of view. They are drivers, cattlemen, mulemen, carpenters, coopers, and masons, as it were in embryo. Their genius and strength rises and ripens with their years, as they are made emulous by proper treatment. It argues then what that kind of treatment should be, to promote with success so good a design. Even in common life, throughout civilized Europe, the welfare of the child is the grand object of the parent. The owner and the over-seer of those valuable shoots should act the part of a parent, fosterer, and protector, looking on them as the future prop and support of the property. How pleasing, how gratifying, how replete with humanity it is to see a swarm of healthy, active, cheerful, pliant, straight, handsome creole negro boys and girls going to, and return-ing from the puerile field work allotted to them, clean and free from disease or blemish.

16

It forms one of the best traits in an overseer's character, to have and preserve such
under his charge. Negro children, after
they pass five or six years of age, if free
from the yaws, or other scrophula, and are
healthy, should be taken from the nurse
in the negro houses, and put under the
tuition of the driveress, who has the conducting of the weeding gang. It is an
unquestionable evil to leave them there
after they come to that age, as they imbibe, by remaining there, a tendency to
idle, pernicious habits. When they can be
any way useful, it is best to send them
with those of their own age, to associate
together in industrious habits; not to over-
act any part with them, but by degrees to
conform them to the minor field-work. A
wide expanse (more or less) of young plant
canes present themselves to the sight, tacitly
calling, by their appearance, the helping,
nourishing hand of man, to aid them in
their growth, by plucking the unwholesome
weeds and grass from among them, to draw

round them the parent earth, the fostering manure, with the tenderness their infancy demands. In general it is found,
that the supple hand of the negro child is
best calculated to extract the weeds and
grass ; and the addition of a small hoe,
used with caution, draws the mold to their
support. A piece of young plant canes,
cleaned and molded by a gang of negro
children, has generally a more healthy,
even appearance, than if dressed by able
people, because they are more light and
cautious in going through it. Few break-
ages take place, and the earth is not trodden by too heavy a body, into a hard
contexture ; a great injury to young canes.
An experienced negro woman in all manner of field work, should be selected to
superintend, instruct, and govern this gang
of pupils, armed with a pliant, serviceable
twig, more to create dread, than inflict
chastisement. I should prefer a woman who
had been the mother of, and reared a number of healthy children of her own, to a

sterile creature, whose mind often partakes of the disposition of her body; who is stern without command, fractious and severe, with an indifference to impart instruction. Each child should be provided with a light small hoe, with a proportionate handle to it well fixed. These little implements should always be ground for them, when out of order, by a carpenter or cooper, and kept wedged; they should be furnished with a small knife, and small basket each, calculated to carry dung. They should be accustomed, in planting time, with those baskets to attend the great gang, and throw dung before them in the cane-holes, which they can do expertly; and by this they will be taught to observe the mode of planting, and putting the cane in the ground. They should be encouraged when they do their work well, and when the sun is unusually powerful, with a drink made of water, sugar, and lime-juice, such being cooling and wholesome for them. They should be minutely examined and cleaned from chegoes; their heads and bodies from itch or scrophula; which last, when discovered, they should immediately be put under the care of the hothouse doctor, physicked and rubbed with proper ointment, and not sent to work till they are cured. Their cleanliness should be exemplary, their meals always strengthened with a small quantity of salt pork or fish, and some kind of garden-stuff, such as peas or beans. And I beg leave here pointedly to remark, (which I hope gentlemen of the old school will excuse me for, as it is an old practice,) that on no account should these, or any children, be sent to gather hogmeat or cut grass, or carry hogmeat or grass to the overseer's hog-stye or mule-pen. The reason for my thus formally declaring against this practice is, that in searching for, and gathering hog-meat, and cutting grass, they are obliged to go a considerable distance to gather it, through wild bushes and woods. They are incautious in thus rambling, often getting thereby bad bruises, hurts and wounds,

which turn out to be incurable sores, ever
after rendering them infirm, perhaps decrepid, and pitiable creatures to the sight.
Respecting cutting of grass, the evil is as
much to be dreaded; for those young creatures are flighty, and unsteady in using a
bill or a knife, and by some mischance
may give themselves horrible cuts, equally
as unfortunate, and to be guarded against.
Besides, an old negro or two will always be
found, who can provide a sufficiency of
these things, and old weakly mules to bring
them home. When any of these children
become twelve years old, and are healthy,
they are fit subjects to be drafted into the
second gang, going on thus progressively
from one gang to the other, till they are
incorporated with the great gang, or most
effective veteran corps of the estate. Crab-
yaws they are subject to, as well as able
negroes ; a species of bonions, affecting the
soles and sides of the feet, having a kernel
deeply rooted, (and perhaps attended with
an abscess) which requires caustic to eradi-

cate, and are obstinate to effect a cure of;
but patience must be called in as an auxili-
ary remedy. The slave must be confined,
his foot clothed with a kind of sandal,
(called in Jamaica a sandpatta) the caustic
applied, the foot kept clean with warm
water and a mixture of goulard, and not
turned out to work till the cure is effected,
and the parts made callous against future
impressions of the kind; I have no doubt
but these rules will be found adviseable,
reducible to practice, and that I shall not
incur the displeasure, envy, or ridicule of
any person, by propounding them.

Cattle and. mule-boys. — This description
of working slaves, (as they are termed in
Jamaica,) should be taken from the great
or second gangs, as found most applicable
to that kind of employment. Youths from
twelve to twenty years of age, and old
negroes (especially Africans) should never
be put to such a task, or taken to be trained
up to it if possible, being too heavy, generally stupid, hard to be subdued to it, rather

ungovernable, when from under the drivers auspices, unhandy, and liable to make the cattle and mules suffer under them. Take then the tractable, docile youth, of creole birth, for most of them know how both to lead and yoke cattle, and ride and tackle mules. When work of that kind is wanted, the head cattle and mule-man has the charge and direction of them. Each should be provided with a well-appointed whip, that may inflict a smart, but not a cruel stripe on a beast, whom they should never be suffered to maltreat. They should never be allowed to ride mules up a hill. They should know, and be instructed in the best method, of dressing the stock for bruises or wounds. They should never have an excuse, that they are unprovided with a sufficiency of good pads, fine trash, and well-made ropes. Each mule-boy should be appointed to the precise mules he is to work ; his spells of mules, as to their names, should be told him. They must be made to tackle them well, spell them regularly, rub them down, and if bruised or hurt by improper working, be punished for such misdemeanor. They should never be allowed to take home the ropes, or mule-pads, to the negro houses, (which is often done,) they being generally giddy, and negligent of them, and inclined to steal them from one another. They should be constrained to deposit them, in a safe convenient shed, built for the purpose. They should be strictly made to keep dry pads next the backs of the beasts, to prevent them from galling or giving colds or spasms to the animals. They should keep them if possible to the custom, (when a number of mules are to be worked, in carrying canes and copper-wood,) of going in a regular gang together, that the head mule-man may always have them under his eye, to prevent accidents. These are the requisites I presume most necessary in training and governing the cattle and mule-boys in their duty. The feeding of the cattle and mules is superintended by the head cattle and mule-

man, assisted by a deputy, who are all
directed in this essential point by the
overseer.

Watchmen, invalids, and superannuated.—
Watchmen on an estate or a property in the
West Indies, which are stationed on the
lines, cane pieces and provision grounds,
are slaves in the light of sentinels and
piquets. It is indispensably necessary to
have such a force in existence : they act
an important part, by their vigilance, to
prevent the trespassing of cattle, or the
depredations of thieves; to repair broken
fences in their neighbourhood, make baskets,
pads, pegs, ropes, &c. As some slaves begin to decline by inevitable old age, infirmity, or disability to stand the more
heavy laborious, field work, they should be
allotted to those kinds of occupations which
do not bear hard upon them. Something
they should always have to do, to keep
their minds employed, and their bodies in
easy activity. This kind of duty comes
within their capacity. An intelligent, trusty

man of this sort, should always be stationed
on the line of the estate, and another in
the negro, and white people's provision
grounds, to guard with care those tempting
places. The watchman who guards the lines,
should be made to conform to the practice
of daily bringing a quantity of bark, fit to
make ropes, to the overseer's house, make
his report of the state of affairs in that
quarter, and regularly, without much delay,
return to his post. The head watchman
should be particularly attentive, that these
piquets are inflexible and steady ; comfortable huts they should always have ; a sharp,
active dog for their companion, and armed
and provided each with good cutlasses,
bills, and knives. No disparagement
should be shewn to them, on account of
their growing old or infirm : these are the
dispensations of Providence, which no
human art can control. Their real wants
and comforts should be attended to, although they do not require, (from the
nature of their employment), so much salt

provision, or frequent change of clothing, as the able field negroes, who are exposed to precarious, inclement weather and hard toil. Every watchman, no matter where placed on an estate, should always have a number of rat springes set in various directions, especially among the cane and corn pieces, which they should be subtle in fixing, diligent in daily examining, and those within the sphere of cane cultivation, ought to be made to produce them every morning at the overseer's house. Nothing is so destructive to a piece of ripening canes, as this gnawing destructive little animal ; no creature, I believe, in the scale of quadrupeds, is more prolific, or more cunning to evade pursuit, retreating to its subterraneous, mazy habitation upon the smallest alarm. It is wary of the snare, yet unceasingly voracious. All methods should then be attempted to catch, destroy, and extirpate them with safety ; men, dogs, deadly mixtures, to entice them ; even fire and water should sometimes be called in, to assist in this undertaking. For it is not the great quantity they eat, but their roving propensity in running from cane to cane, from one piece of corn to another, nibbling and biting to the very core almost every thing within their reach. For wherever they insert their teeth, that, and the adjacent part of the cane becomes sour, discoloured, and gangrened, the vital juices are stagnated, great part of the cane is unfit to make sugar, and consequently the crop is much diminished. I am inclined to hold forth bribes and rewards, for the greatest number taken ; a small quantity of rum or salt pork to each watchman, who may catch in the course of the week so many dozen of rats, keeping a daily book of account. Yet he must be cautious, that the same rats are not brought twice by the watchmen. The remaining watchmen should be scattered at proper places over the estate, where most vulnerable, and liable to attract, by the alluring sight of ripening cane and corn the prowling thief, or the browsing

beast. Repairing of fences, pad, and basket-
making, &c. they should occasionally be
employed at. The head watchman's business is to superintend and direct them.

The supernumerary invalids and superannuated persons, who can do any slight
work, together with such middle-aged
slaves as are afflicted with asthma, bone-
ache, or other disorders which require
occasional rest, should be put under the
direction of a sensible negro of their own
sort, and occupied in planting and cleaning
quick-fences, either round the cane or
grass pieces. Though much cannot be
expected from them, yet it is best to keep
them at some employment ; and such work
is easy and of utility, sparing the necessity of drawing off more able people to do
it. Nothing is more strikingly pleasing
to the eye, than well-planted, well-kept
fences ; they preserve, in a great measure,
by their encircling, binding protection, the
young plants, and the rich harvest of canes,
which a kind Providence and produc-

tive nature, with the laborious art of man, brings to perfection.

Young children and infants. — It is usually the wish of the female slaves, when they become mothers, to keep the infants sucking to an extraordinary or excessive time, sometimes for three years; with the two-fold view of making the child strong, and having loitering, idle time to spend. The latter motive, I believe, is the most predominant. But whatever it may be, it is a bad practice, and injurious to the woman and the child. It reduces the woman to a state of weakness, and barren-ness, and makes her prone to idleness and disaffection to work. The child becomes accustomed to too much tenderness, un-suitable to its station, giving it a fretful longing for the mother, and her scanty milk, engendering disease, and what is worse than all, often (though secretly) giving it a growing liking for the hateful, fatal habit of eating dirt, than which nothing is more horribly disgusting, nothing more to be dreaded, nothing exhibiting a more heart-rending, ghastly spectacle, than a negro child possessed of this malady. Such is the craving appetite for this abominable custom, that few, either children or adults, can be broken of it, when once they begin to taste and swallow its insidious, slow poison. For if by incessant care, watchfulness, or keeping them about the dwelling-house, giving them abundance of the best nourishing food, stomachic medicines, and kind treatment, it is possible to counteract the effects and habit of it for some time, the creature will be found wistfully and irresistibly to steal an opportunity of procuring and swallowing the deadly substance. The symptoms arising from it are a shortness of breathing, almost perpetual languor, irregular throbbing, weak pulse, a horrid cadaverous aspect, the lips and whites of the eyes a deadly pale, (the sure signs of malady in the negro) the tongue thickly covered with scurf, violent palpitation of the heart, inordinate swelled belly, the legs and arms reduced in size and muscle, the whole appearance of the body becomes a dirty yellow, the flesh a quivering, pellucid jelly. The creature sinks into total indifference, insensible to every thing around it, till death at last declares his victory in its dissolution. This is no exaggerated account of the effects and termination of this vile and hateful propensity. As I said before, the mothers of sucking children should be allowed a pint of flour or meal, besides sugar weekly from the store, as those children not only require additional nutriment, but are inclined frequently to laxative habits of body, which fresh flour or corn-meal corrects. I would never (except sickness intervenes) leave a child more than fourteen months sucking, but generally no more than twelve months. During that period it should undergo inoculation for the cow or small-pock; the former in preference to the latter. When well of this disease, and having arrived at the before-mentioned age, the child should be weaned, taken from the mother, and put under the care of a well-disposed orderly matron, whose particular province should be to watch, clean, feed, extract chegoes from its feet, hands, &c., and present the children at the overseer's house before him every day, where there should be a nourishing pot of soup, with boiled roots, and vegetables, prepared, and divided with impartial distribution to each child; once a month worm medicines should be administered to them, and a dose or two of salts or castor oil.   When the children are three years old, they should be put under the care of another well-disposed old woman, who should follow the routine prescribed to the former matron, as to keeping them clean. She should keep them from three to five years old, in a little playful gang about the works, so that in any bad weather, they could soon seek shelter under the different sheds and stokeholes.   Each child should have a little basket, and be made somewhat useful by gathering up fallen trash and leaves, and pulling up young weeds, so as to keep them stirring, and out of the way of harm. These children likewise should have a plentiful pot of soup, with vegetables boiled for them every day, distributed to them respectively, before the overseer or bookkeeper, with a wine glass of acidulated sugar beverage, and a taste of good rum to each, as an enlivener. Their minds should always be kept cheerful, and the parents' fears allayed, by every attention to their growing welfare. The younger children that have been weaned, together with the weeding gang, should have worm medicines every month. The practice of giving cabbage bark to such children as a vermifuge, (an old custom,) is pregnant with danger. It is a native of the woods in Jamaica, the coat of a certain tree, though not of the beautiful tree bear-ing the cabbage on the top of its stem. And although not unpleasant to the taste, yet it is deleterious, so that great caution is necessary in giving the dose, and appor-

21

tioning it to the age and strength of the
patient. It is powerful in its effects, more
by its dreadful deadly qualities, than for
expelling or eradicating worms. A much
safer and more effectual remedy in case of
worms, or to be given periodically to children, is the cowitch taken internally. It is
likewise indigenous in Jamaica. It grows
upon a creeping, spreading vine, in some
retired dell or glade, generally where it
meets support by adhering to underwood.
Pods of it hang in clusters on the vine,
which are covered by a fine, brown furry
spicula, of the most acute, subtle nature,
yet perfectly safe, when mixed with honey,
thick sirup or molasses. A certain portion
of it, what may be scraped off six or eight
pods, to two quarts of sweets, will be sufficient to give to thirty or forty children,
with efficacy and safety. This dose should
be repeated the following day, and after
that, some glauber salts or castor oil should
be given to each child the next day, to
clear the bowels. Wonderful is the effect of

it in dislodging the clinging worms from the stomach and bowels. Its tormenting spicula adhering to, and insinuating itself into them, they drop their leechy hold, descend to the lower intestines, there cling in writhing agony together, and are expelled by the power of the cowitch in half lifeless, and dead multitudes from the body of the patient. The child feels no unpleasant effects from taking the cowitch internally, when well prepared, by its being mixed with the sweets, till the spicula is separated, and appears like fine, thin small hairs, through the honey or sirup. Neither is it any way dangerous thus taken. The only caution necessary is, to prevent the child from putting its hand to its mouth while receiving the dose, and hinder any of it from falling on the skin, which can be easily done, by placing a cloth over the neck and breast of the child. But some of the children are so good and tractable, that they require nothing of the kind, but open their mouth, and with freedom swallow it. If any hap-

pens to fall on the skin, some lime juice
and a little water will soon clear it away.

The treatment of children afflicted with
the yaws, and likewise old people, as it is
a disease which has tried the skill of the
faculty with little success, I cannot presume to say much concerning its mode
of treatment or cure. Time, and, I believe,
the strength of a good constitution, may
work this desirable end, or partially allevi-
ate or remove it. In the middle-aged and
old it is terribly obstinate. Its nauseous
and loathsome appearance, its frightful ravages, its twitching pains, extending to the
very marrow, brings with it a deformity of
bone and flesh that strikes horror. No
wonder then that tremulous fear of such
contagion will make any one fall back with
frightful timidity, and sometimes leave the
afflicted wretch at a distance, within the
circle of a provision piece, to sustain life,
and let nature perform the rest. Children
are more able to recover from this evil than
elderly people. Cleanliness, simple, nutri-

tious diet, without meat, or salt animal food, is a regimen to be observed with children in this case; alterative medicines, cleanliness, and the same kind of diet with the middle-aged and old. A commodious hut, at a distance from other habitations, should be set up for such patients. The children should not be allowed to associate with elderly people so diseased, as the rancour of the disease in the old may add to the infection, and prolong the cure of the young. Bathing, in a sun-warmed shallow stream, will purge the skin and pores of impurities, give suppleness to the stiffened limbs, banish languor and drowsiness, and may be the means, in progress of time, (especially with the young,) of undermining the disorder, and restoring long-wished-for health to the desponding and afflicted. The younger a child takes this disorder, after it is weaned, the better. The sooner we find the cure effected, and the constitution relieved; and, having got over this disorder, the small-pox, the meazles, and the whooping-cough,

the negro child has passed through the
diseases attendant, and incidental to its
youth; the parent is rejoiced, the overseer
and owner are confident they have a healthy,
promising, valuable subject upon their list,
and little to fear, except what may pre-
cariously happen.

I hope this account of the treatment I
presume best calculated to manage slaves,
may be found acceptable, of easy acquisition, no way derogatory to the more
refined, or better formed opinion of others;
and find its way for adoption, with those
interested in West India capital.

CHAP. III.

CHOICE AND TREATMENT OF STOCK.

THE choice of stock (such as cattle and mules), either for work or breeding, is a leading feature in the principle of good plantership. Much depends upon it. Much is expected from an effective force of well made, strong, healthy stock of this description, or a succession (when wanted) produced from prime cows and mares. The crop is to be taken off the field by them, brought to the mill, and ground there perhaps by these very cattle and mules; carried to the wharf many miles distant, timbers and copperwood brought to the works by them, and manure produced and made from them equal to what may be required; and that with such celerity and safety, that these things not only may be done with due despatch, but the stock comparatively be in good condition. The steer or spayed heifer, before they are too old, or too much reduced, should be turned to the fattening pasture, and sold when good meat to the butcher, thereby sustaining but a trifling loss in the prime cost of the beast, and having some years of their work and manure for the care and feeding of them. The kind of pasture the stock has been bred and reared upon should be looked to. In my opinion, those of a good breed, which have been brought up on well kept common pasture or savanna grass, are much preferable for work, than those which are reared on artificial guinea grass. They are found hardier, their flesh more firm and compact, more docile to be broke to work, less liable to fall off in flesh while at work, more easy to be recovered and restored to health and flesh when reduced, and their hoofs more flinty, tougher, and better able to endure travelling over stony river-course roads. Their meat, when fat,

K

sweeter, better, and weighs heavier. To
these qualifications may be added, that
they are generally a few pounds cheaper.
The same observations hold good respecting mules in a great measure ; whereas the
cattle and mules brought up on Guinea
grass are more tender, bloated, liable to
tire upon any pinch of work, are often
stubborn, restive, and lazy, soon lose their
frothy flesh, are difficult to regain it ; heavy
in their tread, with soft pervious hoofs,
which often split, and contain deep-seated
crab-yaws and ground-itch. There are, indeed, multitudes of fine serviceable cattle
and mules taken off of Guinea grass pens,
tongue steers especially. The proprietors
and overseers of these inclosures take great
pains to have their cows and mares crossed,
almost every two years, by young bulls and
jacks of the best breed, sparing no cost in
the attainment. They are so pampered by
frequent change of pasture, and ranging of
extensive runs, that they attract by their
bulky, plump, sleek appearance, the anxious

purchaser, who is in need of stock for immediate work.   Yet with all their polished,
desirable looks, they have not the stability
of those that have been bred on, and taken
off common pasture, when their breeding
has been taken as much pains with as those
bred upon guinea grass pastures.

The next thing to be considered, is the
form of cattle and mules for the particular
work they are designed for.   The steer
and spayed heifer for work should be firm,
active, and straight in their limbs; straight-
backed, their hoofs should be close, compact,
and of a middle size; their chests broad
and capacious, with a full muscular neck,
light neat head, with straight full horns;
their eyes clear and sprightly, but not
treacherous or wally; great girth of ribs,
especially near the shoulders; the shoulder
large, and well knit to the chest, neck, and
ribs; close and full in the loins; sturdy, yet
active in their hind legs; with small ears
of quick perception in hearing; with no
warts, crab-yaws, or ticks.   Such are the

requisites I presume to set forth, as forming
the bodily abilities of the working steer
and spayed heifer.   I shall now take the
liberty of reflecting a little on an old
custom, much acted upon in Jamaica, which
is, the bigotted pertinacity (if I may so
call it), of refusing to purchase some working cattle on account of their color; and
often choosing weak, deformed, and ineli-
gible stock, because they are of such a
color.   Superstition carries people a great
way out of the reasonable track.   The
ignorant, credulous slave may pretend that
something ominous will attend, some misfortune will follow, buying cattle of a
certain color.   It consists not only in their
own barbarous, ignorant notions, but in
their fondness for a certain colored beast
themselves.   But for a proprietor or overseer to be thus guided, thus predisposed to
cast away the best-made steer, because he
is not brindle, red, or black, is only to
thwart his best interests, and bring losses
and disappointments on himself.   Even in

the choice of tongue steers, so much to be depended on for their strength, steadiness, and size, they are squeamishly captious in this point, and will rove from pen to pen, in search of cattle to answer their favourite colors, spending their time, leaving good serviceable cattle unbought, and perhaps purchasing and bringing home with them, washy untractable stock, which will not stand the trial of a crop. I will not pretend to assert, that cattle of the regular colour of brindle, red, or black, may not be excellent; and when the qualities of strength, symmetry, youth, and docility are united, they are indeed admirable. I only wish to guard some people against the prejudice of color in choosing cattle, and committing a crime against good judgment, in the selection and appointment of steers or spayed heifers for work, and allowing the butcher, by this oversight, to kill thousands of good, sturdy, efficient cattle in the course of the year.

In order to entice nature to produce

cattle of regular colors, such as brindle, red, or black, where a number of breeding stock are to be kept up, for the planter to draw his working stock from, I would propose to make choice of young well-made bulls, and prime well-made three year old heifers of those colors. We generally find nature inclines, to a continuance of the color of the parent beast. Sometimes she is sportive, though not the less kind and valuable in her favours, bestowing beauty by varied colors in the calf. Why then reject the offers of her bounty, why cast a slur on her best efforts, by spontaneously giving well-made stock of brilliant varied hues, which are treated with scorn and contempt, when assigned to, and mangled by the butcher in their prime, before a trial is given to the efforts of their labour. Having the option of the most approved color, the make of the bull and heifer comes principally to be noticed. The bull should rather be long-sided, of massy, well knit, active, straight limbs, have an extensive wide chest, straight and broad back, till within a few inches of the verge of the shoulder, then the back should rise gradually, with great strength of muscular flesh to the contact of the shoulder and neck, exhibiting power in those parts. The neck should be of a middling length, very thick, sinewy, a little bowed, and conjoined to the back, shoulders, and head with freedom. The head not heavy, clear and sprightly eyes, but not wally; the horns springing in a gradual curve from the head, short, light, and spiral; small acute ears, the hind quarters plump and sturdy, with close, full loins, and the hoofs middle sized, close and hard. He should be amorous and fecund, but not ferocious.

The heifer for breeding, should be tall, but not long-legged; her height should be included in her depth of shoulder, girth of rib and barrel, and large buttock. She should have neat active legs, chest large and full, straight back, small head, and moderate well-shaped horns, small acute ears, full, sprightly, clear eyes, but not wally, thick pliable muscular neck, broad full rump, and hind quarters; she should be wide behind, her paps or spins, at a good distance from each other, her udder plump, not skinny or stiff, and capable of considerable distension. Both bull and heifer should be free from the evil excrescences called warts, because if once their blood is infected with this disease, they are not fit to breed from, the cow seldom rearing a strong, healthy calf, and the disorder becomes hereditary.

Respecting mules best calculated for work, whether Spanish or Creole, their color is not much attended to. Indeed little variation occurs in that particular in this animal. It is generally a dark brown, a dun, or mouse colour, sometimes grey and black; superstitious connoisseurs do not dwell much on choice of color here, though they might with as much reason form their objections. I would choose either for draft or back carriage, the young truss-made mule, not too tall, with stout, active, well-appointed limbs, small head, straight visage, quick, clear, sharp eye without blemish, light-necked, sinewy, and a little bowed, large chest, deep strong shoulder, straight and rather short backed, close loined, wide behind, not cathammed or sprawling in their gaits, small, hard, and black hoofs, light pendant main and tail, with small sharp ears, no ticks, or swelled joints, diseased fetlocks, or blemishes. These are the qualifications, I think, when combined, that will turn out, and ensure a good, serviceable, working mule.

Now for the model of the mares and jack to produce such from, if nature is propitious in permitting it. Middle-aged mares, if healthy and well-made, of a good breed, &c., will do as well, and if not better to breed mules from, than young mares. But I will here premise, that I by no means approve of breeding animals of this kind from old, infirm, weakly, disordered, blind or decrepit mares. This is greatly to be lamented, and is too much practised. For, sooner or later, the misfortune of buying stock produced from such beasts, will fall on the owner or purchaser. Whether old, weak, or disordered, the evil lies dormant in the mule for some time, and unexpectedly will break out. Neither is it sound policy in the pen-keeper, who is to get his livelihood by keeping breeding stock of this sort. For a great number of the mules dropped from mares of this description, turn out unfortunate, the dams sometimes not being able to rear them, and if they do, they are a symbol in general of what they sprung from, being weakly, ill-shaped, apparently half starved; and after a great deal of pains taken with them, scarcely pay the owner for the grass they consume, and very often are sold for half price to some stroller, or left on their hands, to be an ornament to a well stocked pen, or rather an ugly disparagement of it. Young mares are too timidly coy, reluctant, restive, and shy of the jack, which generally terminates by their being cruelly bit, or the jack severely maimed, else the groom has uncommon trouble with them. But it sometimes happens, that people are so wise as first to let the young mare to the jack, to prepare her for the future embraces of the horse. This unnatural practice is attempted to be defended, on the score of making them more capacious in their genitals, and enlarging the sphere of their abdomen. But it must be remembered or understood, that the generative parts of a prime jackass, are as large as those of a horse, and when the mule cub is just dropt by the mare, it is as large as a foal that is just born.

The mare to breed mules from, should not be more than fourteen hands high, nor less than twelve. She should have a small well-shaped face and head, small upright sharp ears, fine, clear, and full eyes, well pupilled, straight, firm, and neat limbs, no way cathammed; with small, black, hard hoofs, full, wide, prominent chest; slender,

but muscular neck, a little bowed ; strong
deep shoulder; rather a short body; straight,
fair back ; large barrel ; close between the
hip and short ribs ; large round buttock,
wide behind, with free, easy, bounding
gait ; a temper no way irascible ; gentle
and free from tricks, without mange or
spavin. The jack should be as large
an animal of that kind as can be procured, but proportionate in his limbs. It
is said those of an iron grey colour produce hardy cubs, but that is doubtful.
Spanish or Maltese jacks, which have been
imported into Jamaica at a great expence,
have turned out well, producing excellent
stock ; but they are often very old when
they arrive, bruised, battered, and ignorantly taken care of in so long a passage :
emaciated, half-dead creatures, that require
the utmost care to recover them, and bring
them round. Months often elapse with
patient expectation, before any one can
venture to bring them in contact with the
mare, impotently, yet viciously striving to

generate. A jack should be ten, eleven, or twelve hands high ; his body of moderate length ; his head and joles in proportion with his neck ; his neck thick, of great strength, and rather long; his ears not heavy, yet long, sounding well, and both they and his mouth flippant; his mouth small, well furnished with good teeth, especially the grinders ; straight, smooth, easy back; neat, active, strong limbs, standing sturdy, yet nimble ; large chest, close-loined, round plump buttock. The breeding jack should either be stabled, or put into a close pasture, with high, firm walls and gates to it. They, or he, should be regularly corned once a day at least ; should have pure water to drink, and not suffered to cover more than one mare daily. The mares should be put to him in season, and attended by an experienced groom. A proper covering pit should be made for the mare to stand in, with a surmounting stage for the jack to stand on. They should be daily taken and led out to exercise, kept well cleaned, and by no means allowed to stay out in bad weather, but comfortably stabled, foddered and littered. No other jacks or stallions should be suffered to come close to him, to prevent the mischievous effects of their savage, cruel quarrels. This is the specimen of a mare and jack, that I humbly beg leave to propose as the fittest to breed from, to produce a stock of working mules.

I come now to lay before my reader, the best mode I think should be adopted for the feeding and treatment of working and breeding stock, belonging to an estate in Jamaica. Every estate or coffee plantation should be provided with guinea grass inclosures, independent of or separated from the common pastures, cane pieces, coffee pieces, or provision grounds, to answer both as nurseries for reduced, lame, or fattening stock, and to draw provender from, for the mule stubble, and cattle pens. These pastures or guinea grass pieces, should never be eaten down so bare, but that they could

recover, and present another sufficient
growth of grass in six weeks or two months.
It would be better to have a number of
small inclosures of five or six acres each,
than very large ones, so that the cattle
may be changed frequently, the grass not
much trodden upon, the cattle kept well
filled, the flesh they have collected thereby
not let to dwindle or be lost, and the pastures have a sufficiency of water in each, or
somewhere contiguous to them. The overseer, the head cattle and mule-man, should
not fail to pay attention to this, and in crop
time, when the head cattle and mule-man
may be working stock, the overseer engaged in a variety of business, and not
able to pay much attention to the grazing
cattle, one of the subordinate young white
men (the bookkeeper) should superintend
this duty. According as any of the cattle
or mules become reduced, thin or lame,
they should be first minutely examined,
cleaned of ticks, their bruises and sores
dressed, and then turned into one of these

inclosures, and daily dressed, till their sores are well, and their skin sound. They should be replaced by such cattle and mules of the working class, as may be then, or from time to time, found sufficiently recovered to be sent to work, as were grazing there, for the benefit of their health. Breeding cows and young unbroke cattle, as they undergo no work, and are intended to supply a succession of hardy stock, and have all day to feed and range over the pastures, should be penned at night, on one of the worn-out cane pieces, separate from the working stock, which pen should be well secured, littered, and provendered with plenty of guinea grass or long cane tops. If the weather is very rainy, they should be turned into a close pasture by themselves. A great advantage arises to an estate, by penning the breeding stock on poor worn-out cane pieces. They make abundance of fine manure on the spot, and save the trouble, delay, and expence of carrying it there. The urine sinks deep

into the ground, restores in a great measure the expiring stamina of the earth ; and
the breeding cows with their calves, and
young stock, by being thus used to penning, forget the wildness of their nature in
that country, are kind, docile, and easily
catched to dress or milk.   The young stock,
as they come of an age fit to work, are with
little trouble broke or trained to it.   I
would not pen or stable the breeding mares,
and young unbroke mules, except in very
bad weather, and then in a covered place ;
because, when stabled in that country, they
are very near each other, huddled together,
become restive, vicious, liable to kick
and bite, greedy to eat what may be in
the rack and manger, thereby excluding
many from any benefit of it, producing
often abortions, which reduces the mare for
months ; or perhaps a mule is turned out
in the morning with a broken leg or thigh.
In bad weather, to prevent cramps, colds,
starvation by cold, staggers, &c. (which
cattle are so liable to,) penning or stab-

L

ling should be ventured upon. I would at all times (except in very bad weather) especially in crop time, pen the whole of the working cattle in one or other of the poor worn out cane pieces, or thrown up land, as then plenty of long cane tops can be had, with guinea grass for provender, which will make abundance of manure on the spot. The pens should be well fed, with plenty of guinea grass and cane tops mixed; and as they are made and composed of mortice posts with rails, they should be moved every eight or ten days to another meagre spot, till the manuring of such cane piece is in a great measure complete. As those cattle are regularly spelled, they have a good portion of time to graze, and when penned plenty of herbage and tops for the night to eat, which makes them drop much dung, keep their flesh, and have a hearty sleek appearance. But this must be observed, that no reduced cattle should be penned, but as soon as they shew symptoms of weakness, poverty, &c.

they should be consigned over to the guinea-grass pasture till well, strong, and in full flesh.

It is a received opinion in Jamaica (which is invariably followed), that the calf should be allowed to suck and follow the cow till it is nearly twelve months old, or as it is commonly called in England and Ireland, reared at the cow's foot. I have known them to be permitted this indulgence, even to within two or three months of the cow calving again; which they do on the principle of making the calf strong, and not stinting its growth, forgetting that most of the calves reared in England and Ireland are uniformly taken from the cow shortly after they are dropt, penned up, and stall-fed with new milk, till they can graze, and the cow be regularly twice a-day milked. Yet these cows and calves, thus treated, exhibit a more healthy, vigorous, plump appearance in general, than what are reared in Jamaica. The calf, when grown up, is

bulky and athletic, and the cow much more
docile than those of that island; gives a
greater quantity of milk; seldom falls off
much, (except by excessive milking or
starvation,) and breeds the faster. To this
may be added, that fewer misfortunes happen to the calf by accident or bad weather
in pen-feeding. He is alert, strong, healthy,
fat, and tame when turned out in a grass-
piece. Other circumstances may be started
by the breeder of cattle in Jamaica, that
they would find it difficult to inure the
slaves to such a method, and their prone-
ness to stealing the milk would be a pre-
vention. But the fact is, they have never
taken the trouble, or tried the utility of
such a plan, save now and then in case of
the dam dying, or being lost, they would
attempt to raise the calf by this experiment.
But much oftener the hapless orphan is
consigned to the knife, to give a luxurious
repast. The disorder in this respect, I
believe, arising from the prevalence of
custom, is so rooted that scarcely any argu-

ment would be effectual to wean them from it.

Another misfortune very often happens in Jamaica to young calves, which is, when the cow happens to calve some days before she is discovered, or brings (as they say in that country) her calf out, the horrid putrid maggot fly, so pestiferous there, attacks its tender raw navel, bores into it, and deposits a multitude of embryo maggots, which soon attain life, and eat, penetrate, and corrupt the abdomen with shocking and amazing quickness, so that when the poor staggering innocent is found, it is often so mortified that all the pungent stimulants that can be applied will fail in either killing or extracting the vermin; and the creature dies, a shocking victim of agony, in a short time. It is, therefore, incumbent on those who have the management of them, to have a breeding-book kept, in which should be entered the time the cow goes to the bull, and the time she is expected to calve; and be watchful of that time, not to let her remain out any long space, to catch or imbibe vermin herself or the calf. They should be particularly careful every day to have them dressed with chopped green tobacco, mixed with a little spirit of turpentine and fine white lime, to destroy the vermin, and have the parts washed from any impurities of congealed, corrupt blood, with warm water, instead of lime-juice, and then anointed with a little train oil. I should prefer housing the cows every night, for some time before they are expected to calve.

The working steers and spayed heifers should be classed according to the kind of work they have to do, whether mill or wainage; the light, smart, active young steer and spayed heifer, to be appointed for mill-work, and light cartage about the works or cane-pieces; the strong, large, middle-aged, steady drawing steer, for waggonage to the barquadier; but the mill cattle out of crop, when in good order, and when a large shipment is to be made

with despatch, should be mixed with the
road cattle for assistance, taking care to
place them as middle cattle in the draft,
but neither as tongue or leading cattle. I
would never put the road cattle, intended
for carrying the crop to the wharf, to any
other kind of work, so much depending
upon their veteran, steady efforts, when in
need of them; for when they are imposed
upon, disappointment succeeds, the overseer is vexatiously embarrassed, the head
cattleman incurs blame, (though perhaps
faultless,) the mill cattle brought in, perhaps improvidently, to assist, and all are
reduced in point of strength and condition.
Some work or other is put to a stand, and
a length of time elapses before the cattle
are effectually recovered. I think an attempt should be made in Jamaica to change
the old established custom of binding the
working cattle together with heavy, mon-
strous, wooden yokes and bows while at
work, and that well-stuffed collars, covered
with sound durable leather, would be found

preferable. The collars should have strong
draft rings fixed to them, with all necessary
chains and cross-bars appended; the wain
or waggon should be fitted with strong
shafts, instead of a tongue, that should ply
up and down upon strong iron draft-
hooks, and be fixed to the body of the
waggon, or a draft-bolt. The ease of the
beast in the draft is as much to be attended
to as any other point, to prevent him getting
cross or restive, acquiring a painful, stiff,
swelled neck, or galled shoulder, which
very frequently happens when he is en-
cumbered with those heavy yokes and bows.
He would be more at liberty in the collar
to use his strength, without bounding aside,
to the injury of his driver or fellow-steer,
the side and centre chains preventing him.
Moreover, they would draw more even,
and with greater ease, the heavy carriage,
with its ponderous load, with the aid of
good strong swinging shafts, well fixed with
draft-irons, than by the neutral tongue,
which often shakes the tongue-steers nearly

breathless.   A sufficient number of draft
cattle should always be kept on every estate
to allow of regular spells both for road, cane-
piece, and mill-service.   It is a gross error
not to do so, as the loss in the long run,
by a niggard strength of cattle, is severely
felt by the proprietor.   The road-cattle
should never be worked more than every
other day, whatever less they may be ; and
they should be well fed, and dressed of their
bruises.   Nor should the mill or cane-piece
cattle either, but with this difference, — in
the mill and cane-piece cattle, the former
should be spelled and well fed every three
hours, and the cane-piece cattle every six
hours, paying attention to their bruises.
Any description of working cattle should
never be strained, or forced against their
known strength, which often happens,
through the merciless ardour of the cattle-
boys, and the poor beast is paralysed and
bereft, by such treatment, of all power of
its hind quarters, seldom recovering its
strength, and generally becoming a dead

weight on the property for its support; or after two years of precarious life in the best of pasture, is sold half fat to some neighbouring butcher, for a small compensation. The same caution I will beg leave to give, respecting the overloading of mules. But here the loss is a total one, even sometimes tempting the owner to shoot the creature, to put so wretched an object out of his sight. So parsimonious are some employers, especially resident agents, their memories so defective, or so tedious in granting what is absolutely necessary, that they will both see, and let the working cattle, and mules, on a property, dwindle away more than one half of their usual complement by overwork, old age, casualties, or the like, before they will comply with the repeated requests, and admonitions of the overseer, for a fresh supply; and with a surly rebuke in the end, blaming him for the mortality, perhaps discharging him for it, when their own supineness, craft, or stinginess was the occasion of it. They will give a small spell, perhaps, to sustain nearly the entire of the future work, they drooping likewise by being imposed upon. But where breeding cattle and mares are kept on an estate, this seldom happens: several facts of this kind I have known in Jamaica. One that happened about eleven years ago on an estate, which had a great part of its best plant canes to cut, with some excellent ratoons, to make the crop up. So reserved was the resident agent, so skilful in keeping his mind to himself, so pompous in doing mischief, vainly thinking he was doing good, that the overseer, after months of reiterated application to him, to have a spell or two of young fresh mules brought to help to take off the canes, and save the old mules from premature death, never even once obtained a reply to his entreaties, or a beast to assist him. A great part of the canes were left uncut, a prey to rats, rottenness, topheavy from suckers, and stagnated, and dried up of

their juices. He discharged the overseer, without assigning any reason for so
doing, sent a novice in his place to manage
the estate, discharged him in a few weeks,
and succeeded him by a prodigal overseer ;
and at last, by a variety of management, in
the course of a year abolished, by his mere
sign manual, the studied concerted plan
of the former overseer, (who had established a fine field of canes for a present and a
succeeding crop,) and threw the estate back
in its accustomed, expected crops for years.
One piece containing ten acres of fine plant
canes, the former overseer had partly cut
down, promising three hogsheads of sugar
per acre, not far from the works. This
piece of plant canes, presented to the
astonished eye of the well bred planter the
disfigured appearance of six or seven
growths of canes upon it, besides part of
the high canes upon that cane-piece not
cut down, after a space of four months,
from the commencement of its cutting.
What regard could such an agent have for

the interest of his constituent ? This estate lies in the centre of a well watered vale, in the parish of St. Mary, and is distinguished for its hospitality to strangers, who pass from the south of the island to Rio Nuova Bay, or Salt Gut.

In dry warm months, in Jamaica, the insect called the tick is very abundant, sticking to the cattle, and breeding on them in clumps, burying their heads underneath the skin, drawing and obtaining nourishment, by sucking the blood of the beast, and thus pestering, infecting, and distressing it. They adhere principally to the inside of the ears, and over the body; in horses and mules to the inside of the ears and fundament. When the beast is observed to have them, they can be easily banished, before they get too large, by rubbing the part they cling to with a little train oil, and the next day washing the part with salt-beef pickle, salt and water, or if near the sea, by swimming the cattle in it once a day for some time. The dunder or lees of the liquor still may destroy them. Cattle never look plump or sleeky when possessed by these vermin, therefore they should never be suffered to grow to any size on them, for sometimes they make them look all raw and scabby, from their voracity to feast on the best qualities of their blood. The dysentery, and purging called the scour, often attacks cattle in that island, either from grazing on young unripe grass, or some morbid matter in their intestines. They should in that case be housed for two or three days, have a strong dose or two of glauber salts, mixed with some sweet oil, and the fat of herring pickle. They should have, twice a day, some parched corn given them in a little water, plenty of sound ripe grass to eat, and be comfortably littered at night. The litter, with what grass may be left, should be taken cleanly and carefully away the next day, and put out of the reach of other stock ; for this disorder is infectious.

The proper method of working and

feeding mules, and tackling and relieving them when sick and sore, should always engage the attention of the overseer, or those under whose care they may be. Breaking them to back carriage is easily performed, or to that of draft; yet caution must be used, to have good strong tackling for so doing, and other mules in company. The load should be very moderate for some time, and they should be put to work in the centre of a triple, or three mules, the leading mule inducing them to follow, and the rear one keeping them steady, and free from tricks. In a day or two they will be tolerably gentle and manageable. A principal thing to be attended to is always to have a sufficiency of good, well made straddles, crooks, pads, ropes, and fine trash ready; the straddles to fit the back well over the pads, of good length, and lined with seasoned, tough, light boards to the end, to which should be strongly attached, seasoned, wide, guavee crooks, properly bored, with strong cross sticks wedged thereto. Some hackled plantain trash (but tow would be better) should be strewed thickly over the spine of the mule's back, before the pads are put on, to prevent rubbing and galling. No less than three well-made platted pads should be put on each mule, that has back carriage to undergo. As soon as an under pad be-gins to fall to pieces, or gets wet, it should be replaced by the next pad to it, and a new one got as an overhale. This should never be neglected, else a stubborn sore back will ensue. The pads should be large enough to extend from the hip to the neck, the breadth to the extremity of the ribs. There should never be less than two girths for each mule. They should be platted at least two inches broad, where they are to bind on the belly, be strong and pliant, especially on that part. There should be a strong wanty, of good length, likewise platted as the girths, to each mule; and a well-made halter for each mule, with platted noseband, headstall, and chokestay. Those ropes can be made by a handy negro watchman or invalid, of seasoned bark, found in the woods of Jamaica in great plenty, and a regular sufficient supply kept up at little expence. Mules thus equipped for back carriage, will carry a considerable load, of one hundred and fifty weight of canes, with ease and safety, except the mule-boy, through neglect or villainy, causes some misfortune, for which he should be punished. When the mules are spelled at dinner, or any other time, they should be well rubbed down, their backs examined, and if found swelled, bruised, or galled, immediate application should be had to the requisite dressings for relief. Strong singlings, or low wine to wash them with where swelled, or bruised, should be used, and a little spirits of turpentine, oil-nut leaf, and fine white lime, mashed and mixed together as a plaister, to dress scratches, cuts, or galls with, and the part so affected be touched with train-oil to keep the flies off. If they have bad sores or swellings, they

M

should not be worked till they are well. I would recommend boiled beef pickle, now and then to rub their backs with, as it renders callous and tough, those parts most liable to be affected by friction or weight. Care should be taken, on no account to allow the mule-boys or their drivers to ride them up hill; for such a burthen comes on the foremost mule, added to the struggle, of dragging the followers in his triple, along with him, as tires, or soon breaks his wind. Exertion should be made in the day-time, to have a sufficiency of canes brought to the mill to last all night; and the mules must not be worked late at night if possible; for it is mostly at those unseasonable hours they get bad sores and colds, and may be, as is often the case, stripped of their tackling by the mule-boys, without being rubbed down, or their wounds attended to. The mules which come in from work, either at night or in the day, should always be put into a division of the stable by themselves, where the rack and manger

should be well filled with fresh grass and
cane tops, else the poor hungry animals,
by being huddled together with the rest,
come to short commons, or often nothing
to satisfy their appetite with, every eatable
being devoured by those who were penned
up hours before them. There should be
always four divisions, with full room for
the stock, in a mule stable, and dry grass
or litter of some description, to strew the
bottom of the stable with in the evening;
but this is very seldom done. It will pay
very well for any trouble and expence, by
the manure it produces, which should be
taken clean out every day, and heaped up
in a convenient manure pit. Mules carrying canes to the mill, copper-wood, or
country staves, should always be spelled
every six hours, and abundance of provender kept in the rack and manger for them.
Those which are spelled in the day-time,
after being rubbed down, cooled, and
dressed, should be turned out to graze;
for it gives them great refreshment to have

liberty to tumble and rub themselves, besides that picking fresh herbage is grateful to them.

On no account should the mule-stable be suffered to accumulate a heap of dung; it should be daily cleaned out. The pens being covered in, the heat of the climate, with the warm fume issuing from a number of beasts, is sufficiently to be dreaded, in causing and spreading distempers among them; but that of the accumulated heat, and putrid vapour of a dung heap, in a close mule stable, is pregnant with the most pernicious, sometimes fatal consequences. Their hoofs are kept soft by it, and their blood in a ferment from its noxious sweating qualities. Some of the beasts are more liable to disease than others. Some are not free from it, though apparently looking well. Others have lurking disorders, which are partly discharged by their excrements; making a compound of vile materials to cause pestilence, which when once epidemic, carries off great numbers. I would therefore recommend the utmost cleanliness in a close mule pen, or even in an open one. As I said before, both in and out of crop, the mule stable should be well supplied with wholesome fresh provender ; but in crop time, when heavy laborious work requires stronger nourishment for the beast, plenty of fresh cane tops should be chopped small, so as to fill the manger. These should be strewed over with a small proportion of salt, a good deal of fresh mucous cane skimming thrown in, and if plenty of Indian corn on the estate, a pail or two of it ground, and mixed with the cane tops and skimming. This will keep up the strength of the mules. But care should be taken, that the manger be cleared every day of any remnants of this, for fear of its becoming sour, and causing thereby bellyaches to the beasts. The rack should always be filled with fresh ripe grass, and care should be taken, that the mule tackling be put up every night in a covered place to hang on, and not carried to the negro houses. The disorders of mules are various, but the cure of them is little understood, or only partially known. A book of well approved farriery should be kept on every estate, and the instruments requisite for that profession, such as phleams, syringes, &c. Bellyaches are very frequent with working mules, especially in crop time, which is principally brought on by their either eating or drinking sour cane tops or cane skimmings, or from the crudeness of their provender in general, their natural liking to bite at any thing that has the appearance of an eatable agreeable to them. This protracted spasmodic affection is often so terribly violent, as to cause the death of the beast in a few minutes. They swell to an enormous degree, rolling and groaning in agonising convulsions, till they nearly burst. They shew symptoms of this disorder very soon after being attacked by it. They paw and scrape the ground with one of their fore hoofs ; droop their head nearly to the ground, incline their head often to one side and the other, with seemingly painful solicitude; heave their loins and belly quick, and have a constant inclination to lie down and roll about. When any of these signs are discovered, they should be immediately stripped of their tackling and led out, run smartly about for a few minutes, then copiously bled, and their head tied up high to a strong rail or beam, and drenched with either six or eight ounces of glauber salts, dissolved in a pint of warm water, or six ounces of castor oil, mixed with one hundred and twenty drops of laudanum, half a pint of warm water, with two ounces of common soap dissolved in it, and half a pint of rum. Care should be taken not to let the animal lie down, till the symptoms subside. It would be best to keep it walking about till the drench operates, or the beast is apparently recovered. It should not be put to work for a day or two, but be kept in the stable, to recover from the exhaustion and weakness brought on by the disorder, and have no-

thing but fresh ripe grass to eat, some
ground corn with a little salt in it, but not
much water. Another disorder they are
subject to, is the mumps, which swells
their head and joles frightfully. This likewise may be of serious consequence, if not
taken in time, to prevent the glands of the
throat and lungs being infected. The beast
should be bled, his head wrapped up in a
warm cover, as far as the contact of the
throat ; his joles to the ears rubbed or
washed twice a day with warm fomenta-
tions, melted hogs-lard, bees-wax, and spirits
of turpentine mixed together, and made
warm, till either the swelling goes away, or
suppuration comes on, forming a soft tumour,
which, when ripe, should be lanced to let
out the humour, and kept open by a tent,
in order to discharge the virulent matter
which flows to that part, giving natural
relief to the animal. When the cure is effected by its drying up, and the swelling
disappearing, then the orifice may be closed
and healed up ; the animal should be phy-

sicked, have warm corn-mashes, be kept in a stable apart from other beasts, led out twice a day to exercise, if the weather will permit, and be supplied with soft, fine ripe grass to eat.

The farcy often attacks mules in Jamaica, and is generally brought on by over-heating, the blood becoming surfeited, bad and grumous. It may be occasioned by bad unwholesome diet. It is easily cured if early attention is paid to it, otherwise it will run through the whole system. The button-farcy first appears, by the veins of the legs, thighs, and breast, exhibiting a number of excrescences and tumorous knobs. The animal should be bled two or three times, not profusely, taking frequent notice of the increase and decrease of the disorder. It should be drenched with opening medicine two or three times, taking sulphur bolusses, which may both drive out the disorder, and sweeten the blood. The tip of some of the largest pustules should be taken off with a sharp knife, and a coarse grain of corrosive sublimate introduced into them, and then closed with a little mould candle-grease. The corrosive sublimate will penetrate to the adjacent pustules gradually, and what with small bleedings, sulphur bolusses, occasional physic to clear the bowels, and wholesome nourishing food, the beast will soon declare its recovery to health, by shewing a clear skin, and the arteries, veins, &c. being reduced to their proper state. As this disorder is infectious, it is best not to allow the diseased beast to keep company with others till it is cured. The water-farcy is very obstinate, odious, and often fatal. The whole mass of blood is morbid corruption, which issues from the eyes, ears, nose, and surface of the body. A horrid scrophula spreads over the whole body. Nothing but alteratives, frequent bleedings, and wholesome food, will work a cure. Time will often gain the ascendancy, with those auxiliaries, and restore the creature to health ; but I have known some fine stock to die of this disorder, a shocking

21

emaciated spectacle, of putrid, coagulated matter.

The glanders is another dreadful disease, which mules are frequently attacked with. This is rendered more formidable by the imperfect knowledge which most people have of what may effect its cure, or stop the deadly contagion, which spreads with amazing rapidity, making people panic struck as to its ravaging consequences, or how to stop it. The fundamental cause of this dire disorder is variously assigned, but I believe it is principally brought on by neglected colds, strangles, or mumps, which at last attack the glands of the throat and lungs, pouring through the nostrils a continual stream of thick humour, which at last preys with such virulent effect upon the membranes of the nose, as to rot and disunite them, causing the bones of that organ to drop and fall to pieces, with mortified, putrid, contagious malignancy, and in a day or two putting a period to the life of the ill-fated creature. So epidemic,

it is alleged, is this disease in horses, that
the animal must be removed to a considerable distance, that the very air may not
waft the disease to others. When a con-
firmed glanders is pronounced to have
seized a beast, the death-warrant of shoot-
ing goes forth against it, and the animal
with the distemper is consigned to the
flames; the neighbours are alarmed, the
public cautioned, the very laws of the
island are brought in force to stop the
contagion, by proscribing every beast found
in the public road possessed of it; giving
liberty to any person to put an end to its
life. I have known on one estate two sets
of mules (composed of fifty each) to be
carried off by it. At last the mule-stable,
with all its apparatus, mule-tackling, &c.
were ordered to be burned to ashes, that
no trace of infection might be found. Like
a plague, the disorder did not stop, for
victims, though not so numerous, fell a
prey to it. The scourge at last ceased,
leaving a melancholy fearful impression on

the mind of the manager, and those concerned for the property. A considerate,
benevolent gentleman (who is now no more)
was then resident island-agent for the estate,
and with true philanthropy, he minutely
examined into the cause of such a misfortune, and finding the overseer and white
people guiltless of any fault, supported
them under their anxiety with assurances
of his good-will.   He found the estate
otherwise thriving under their care, and
after giving a fresh supply of mules, he
continued them on the estate; thereby
evincing to the public and the negroes on
the estate, his impartiality in doing justice. A common saying with that lamented gentleman was, that discharging an
overseer, who knew his business, without
sufficient cause, after he had been twelve
months managing a property, was entailing
confusion, and destroying the crops for
more than two years to come — An expression emphatically and truly verified, and
would to God, not practised and followed

up, by his successors, and agents in general. The remedies for this dreadful disorder of the glanders, practised in Jamaica, are bleeding, opening medicines, fumigating the nostrils with tobacco, sulphur, and pungent things, to cause a copious discharge and rowelling under the joles. But I have seldom seen cures effected there with success. The disease either with speedy violence kills the beast, or causes it to linger for a week or two, groaning with inward pain, till its emaciated frame sinks lifeless. I presume to think, that bleeding is the first requisite in promoting a cure; secondly, the beast should be kept warm, and have warm mashes frequently of clean ground corn and young wholesome grass. The fumigating system I would explode altogether, as tormenting the animal without doing any good. It may draw more running or discharge from the nose, but will not clean the nostrils or glands from the corroding, adhesive humour, or reach the seat of the disease. Moreover, it will rather add to the misery of the animal, whose lungs are affected by the distemper, nearly suffocating and depriving it of its already-impaired breathing. Instead of this, I would make use of a good syringe, with a long pipe to it, and with it inject, and wash the beast's nostrils three times a day, with warm water, mixed with a little vinegar and honey, which should be thrown up so far, that the liquid may reach, and drop to the throat, the fountain-head of the disorder. It would cleanse the nostrils of the peccant, corrosive humour, perhaps prevent the humour from turning fetid and green, (the first fatal symptom that presents itself,) and not distress the lungs. I would have a mixture made up of several ounces of balsam capivi, two or three quarts of sharp porter sweetened with molasses, put into a close vessel and shook well, and give the beast a pint of it three times a day. The balsam of capivi is very penetrating and strengthening, the porter antiputrescent, and the molasses of an opening nature;

three things which I presume will do some good and no harm. I would likewise give the beast, for its drink, a decoction of lignumvitæ wood, (which should be made small with a coarse rasp,) and be taken cool. This is all I can recommend for the relief and cure of a disease which has baffled the skill of the farrier and the experienced practitioner.

Other diseases of a minor nature prevail among mules, such as the lampas, botts, &c., the first of which is easily removed, by cutting away with a sharp penknife the fungous flesh that grows between the upper teeth and roof of the mouth, sometimes so as to overhang the teeth, and prevent the beast from eating freely. It is easily separated by a handy person from the natural flesh. Some rub salt on the part after the operation, which I think not requisite, for a little bleeding from the mouth does the animal no harm. It will soon resume its craving for food, and the part heal up. Botts are sometimes troublesome to a beast:

they make it poor and spiritless. When
they are observed, the animal should be
kept up in a stable, and have a bolus, made
of forty grains of calomel, nearly half an
ounce of sulphur, mixed with soap, and
fifty drops of laudanum. In two or three
days after, another bolus, made of 140
grains of jalap, and better than an ounce
of soap, in order to clear the bowels. It
should have a warm mash or two of ground
corn, with good ripe grass, but no cold
water, till the effects of the medicine have
passed off. The staggers now and then attack
mules, for which they should immediately
be bled copiously, wrapped in a warm horse-
cloth, and not be exposed to the sun for
two or three days. The limbs, back and
loins, should be rubbed with warm singlings, or low wines, as it is termed in
Jamaica. The cure is easily attained
by early attention, and the before-mentioned remedies. Having nearly finished
my remarks, on the treatment of cattle and

N

mules, and the best mode that I presume
can be followed with success, as to breeding
and working them, with as few losses and
disappointments, as natural and incidental
causes will allow of, I shall here conclude
my observations, with a transient, but I
hope useful description of a disorder, which
breaks out in mules, called in that country
the pox, so termed by the negroes, and
considered as such by white people.   Whether through ignorance, or by giving facility
to such a term, to make the complaint or
disorder better understood, is a thing I
cannot determine.   This disorder suddenly
appears near the fettocks, and lower joints
of the limbs, or various parts of the legs of
the beast, by a swelling occasioned through
the tumour of a vein or artery, being neither
more or less than a blood spavin, which, by
not being checked in its infancy, extends
higher by degrees, till frequent swellings
present themselves.   At length they burst,
and there issues corrupted blood, trickling
down the limbs, in a disgusting manner,

10*

and causing such intolerable itching to the poor beast, as to induce him to bite and rankle the part shockingly, even to the bone. Yet this disorder is so little studied, and so ignorantly dealt with, that no other remedy is scarcely made use of, to suppress or cure it, but to powder the part that has burst and is laid open, with strong white lime; to give the beast no rest by day, or food at night, it being then tied up to prevent its biting itself; so that the poor animal sometimes becomes a mere skeleton in a short time by such treatment. The white lime adds pungency to the titilation, which is nearly insupportable to it. Thus it is kept and put to sidework, till the disorder wearies, and stops itself for some time. The cure for this disorder is simple, and easily effected, and is no other than laying open with a sharp pen-knife the flesh on each side the tumoured vein, both below and above the part affected, till the vein is sufficiently exposed to see the back part of it. Then having some fine thread, with a good needle prepared, and passing the thread, by the aid of the needle, behind the vein, draw it out by the needle, to the other side of the vein, and tie up the vein with the thread above the tumoured part. The same operation must be performed below the tumoured part, cutting the thread short, after tying the vein. The vein being thus closely tied up, prevents the tumour from extending to other parts. When this is done with no great trouble, the wounds should be closed up, and have plaisters of healing ointment applied to them, and fastened with a bandage, which will soon heal, making the beast as sound as ever. This disorder is brought on by a strain of either an artery, vein, or sinew, and not by any imbibed infection. The beast must be kept in the stable for a few days, with its head tied up, so that it cannot bite the part under cure, and be well fed with plenty of fresh ripe grass and pure water, and if low in flesh, it should have a feed of ground corn daily, and should be exercised during that time, twice a day, to prevent its joints swelling.

CHAP. IV.

BEFORE I proceed to treat of the noble science of planting the sugar-cane in Jamaica, rearing it to maturity, and manufacturing its essential juices, into the staple commodity of sugar and rum, it will naturally be expected, that I should make some observations on the different houses, utensils, &c., used in producing it, avoiding as much unnecessary expense as possible, uniting stability with usefulness, without unmeaning gaudiness, and giving a central position to the works, where plenty of water can be had with convenience. Whether on a level or a hilly estate, the great utility of a central situation to place the manufacturing houses upon, must be apparent to every one interested in such an undertaking; still that situation would be

imperfect, if water, that necessary element,
could not be brought in to aid the works
by its active powers. If a stream of water
does not naturally pass by such a spot, a
course should be levelled for one, from a
source to send down a supply. If such
cannot be obtained, a well or pond should
be sunk, either to draw or collect it from.
But this last is a dernier resort, mostly
found deficient, and of bad quality. A
situation, uniting within itself the blessings
of a plenteous supply of wholesome water,
on a piece of ground sufficiently large to
admit building an extensive set of works,
overseer's house, hospital or hot-house, &c.
with a large mill-yard, and being central
among the surrounding cane cultivation, is
a place most desirable. Having happily
found such a place, a well-contrived plan
of the buildings, their relative, convenient,
and appropriate situations, one to the other,
should be digested, and laid out on a piece
of paper, of a size sufficient to have the
whole delineated upon it. Then having

the materials to go to work with, I would commence with the overseer's house, which should be built compact and convenient, not over roomy; and raised sufficiently high from the foundation, with good masonry work, to admit of suitable stores underneath, to keep all the plantation stores and supplies in. It should be placed so, that all the works can be seen from it, and not far from the boiling-house. The rooms should be all on the same floor, and closely boarded with seasoned stuff. Each white man should have a small bed room to himself, with a glazed sash window on hinges, and a shutter to it. The bed-rooms should be eleven feet by nine each, of which five should be in every overseer's house on a sugar estate, leaving the overseer's room somewhat larger than the book-keepers. A large well-covered piazza, with comfortable glazed windows, (to rise and fall occasionally,) will answer all the purposes of a dining and breakfast-hall, and for walking in. Large centre halls in such houses are of very little use, take up a great deal of room, are very expensive, and make the house large, without any real convenience. A small back piazza, made comfortable by moving blinds with stops, would be proper for the servants. I think every dwelling-house on a plantation, should have a small fire-place in it, with a well-raised chimney, for fire occasionally in damp weather to be made in. It will be wholesome and preservative. The fire-place should be in an extreme angle of the dining piazza, and the overseer's cooking-room, washing-room, &c., should be apart from the house, though not far off, conveniently fitted up, and of a moderate size. The little appendages of a hogsty, fowl-house, &c., to raise small stock in, are easily built at a small expence.

On purpose not to interrupt the view of the works from the overseer's house, I would build the hospital and mule-stable in the rear of it, opposite and parallel to each other, on dry ground, yet not too near to the overseer's house. The hospital should be a strong, commodious building, with wholesome water introduced into it by pipes with turn cocks, a pipe to the male and female common hall, and convalescent room each ; a privy likewise should be made to each hall and convalescent room, having a stream of water running through it, which should be conducted by a sewer, to the mill back water, or descent to carry it off. A good piazza, with moving blinds in the front should be made, for the convalescent slaves to walk in. The body of the building should be comprised of three rooms, twelve feet square each, with a large boarded bedstead, clean plantain mats, and blankets to each slave, one of which rooms should be appropriated to sick male, and the other to sick female slaves. The third room should be for convalescents, and fitted up in the same manner. The hospital doctor's room, should be situated at one end of the piazza. Each room should be well secured with bolts and locks, and the windows iron barred on the outside, and comfortably closed with shutters inside. A fire-place should be in each room, and the house kept clean and often white-washed.

It has been an old custom to confine the delinquents in stocks, set up in the hospital, which I think is a bad practice, as too free an intercourse is given to imbibe disease by it. It would be much better to have a small, strong building made in the centre, between the hospital and mule-stable, of mason work for such a purpose, about twelve feet by nine, with durable stocks fitted up in it, and well secured with strong hinges, iron bars, staples and locks. This little place being thus set apart, and separated from other buildings, will make confinement more irksome and dreaded, and perhaps cause less delinquency. The mule stable as I said before, should be opposite and parallel to the hospital, and of a size according with the utmost number of mules the estate is to have on it, allowing near three feet for each mule, or a mule-stable of one hundred feet long, by

thirty feet clear inside, for every seventy
mules. This stable must be divided from
one end to the other, fitted up with a strong
round staved double rack, and a deep
strong manger to each side. Each side of
the stable should be divided into two parts,
by a wall of cut large stone, well cemented,
so that as the spell mules come in, either at
night or in the day, plenty of provender
may be found in each division. The mules
of one spell, should not interefere with, or
take from those of the other. The stable
should be well paved, having an inclination
of at least nine inches, to throw the urine
or moisture down a channel, sunk a little at
the wall, in the rear of each manger ; four
good strong gates should be made, and
placed at the outside end of each division
of the stable, with good hinges and fasten-
ings. A loft should be made on it, boarded
and divided, one part of which is to hold
corn as a granary, besides new pads, ropes,
baskets, &c., and be well secured with a
door and lock ; and the other part to keep

the spell mules' tackling in, with rails to hang the pads, ropes, and straddles, on. A flight of very strong steps, with hand rails on each side, should be attached to the mule-stable, in order to ascend to the lofts, and the stable should be white-washed inside and outside four times a year, with strong fresh lime, made not too thin.

The mill-house, whether worked by water, wind, cattle, or steam, should be placed as near the boiling-house as the nature of things will admit of, (and the liquor gutter to be as short as possible, with a cover over it, to turn on hinges,) so that the cane heap will not obstruct the passage to either boiling or distilling houses, or sour trash affect the liquor. The mill-house should be built of durable mason work, with strong binding braces in the walls, and sufficiently capacious to admit of the negroes working with ease, and permit the machinery to be taken down and put up occasionally. The cock-pit should always be kept clean, and the cogs free from impediments of trash and dirt. The gudgeons must always be well greased and cool, and the green trash carried away as soon as it is turned out of the rollers. There must be a pipe to convey water to wash the rollers and mill-bed, with a turn-cock fastened to it, and a temporary gutter made from it to the mill-bed. Care should be taken to keep the rollers plumb, well wedged, braced up, and not allowed to be choked with trash. The mill-bed must be furnished with good strong feeding-boards, made of seasoned stuff, to slide to the rollers, with bracing keys, and be well pointed both in front and rear. The splash from the water-wheel must be kept off, (by a feather-edged boarded partition,) from the rollers or mill-bed, and the house always well white-washed.

Two trash-houses should be built at the works on every estate, in a substantial manner, from eighty to one hundred feet long, by forty-five wide each, well roofed, with a cupola the whole length of the roof, to allow the exhalations from the green trash to pass off freely. There should be stone pillars erected at ten feet distance, as high as the wall-plate, to compose the abutments of the building; and in the spaces between the pillars, a firm rail-work should be made, of tough durable wood, (not of bamboos, as is generally the case, which soon get dry and rotten in such places, inducing the negroes to pull them out for fuel,) to keep in the trash. The trash-house should not be entered but at each extreme end, and the trash packed as high as possible, to within a foot of the cupola wall-plate. The firmer the trash is packed, the stronger will be the fuel. Little trash should be permitted to lie about the mill-yard, either to waste such a valuable material, or to make swampy, spungy foot-passages, impeding the carriers, and giving them tender and ground-itched feet. Both about the trash-houses and the works, and their vicinity, whatever bushes, underwood, weeds, or long grass, spring up, should be stocked up by the root, and

the place kept as clean as a bowling-green.
If possible, the trash-houses should be built
on a level with the rest of the works, over
the farther side of the mill back-water, or
upon a level spot separated from the rest
of the works by a small stream of water
brought there for that purpose, over which
should be made a firm gangway, well railed
in, for the trash-carriers to pass to and fro,
with a globe lamp on it, to be lighted in a
dark night. In case the trash-houses catch
fire, being detached from the works by this
stream of water, there is little danger to be
apprehended, and water is convenient for
extinguishing the flames. The slaves, or
any other persons, should never be allowed
to smoke pipes in the trash-houses, or about
the works or cane-pieces. Many dreadful
accidents have happened by such wanton
licence ; the cooper's and carpenter's shop
should likewise be placed and constructed
in such a manner that the spreading of this
destructive element may be prevented ; yet
not so distant but that the overseer can

well perceive from his house how they are
going on there. Every building about the
works of a sugar estate should be shingled
instead of being thatched, and kept free
from the hungry, destructive ant, who by
his mighty, though diminutive efforts, will
level a substantial building to the ground
in a short time. Poisoning by arsenic is
the most expedient mode of getting rid of
them, as the living will feed on the dead,
so that the whole nest, (by devouring one
another,) are thus killed. An overseer
should be as careful of saving and preserving trash as due economy will admit of.
The making of good sugar with despatch is
much to be attributed to its quality and
quantity, for being good and dry, it boils
the liquor quick, throws up its dirty mucous
particles, which is taken off by the skimmer;
and having a quantity of it, prevents the
necessity of detaching the negroes and stock
from other work to procure a requisite supply of fuel.

The boiling and curing-houses should

o

always be proportionate to the number of hogsheads of sugar the estate is capable of making. Those buildings should be composed of the most substantial materials, durable, hard, well seasoned timbers, well put together, and supported by the best mason work. The roof of the boiling-house should be cupolaed from one end to the other; the shingles of the cupola to overhang its wall-plate considerably. The end of the boiling-house that is appointed for the coolers, should have moving blinds, with stops, to admit air and light ; and the other end, where the receivers and syphons are placed, should have an open arch, with a shed on the outside. At this end the chimney is erected, which should rise considerably above the roof, be built of the best fire-bricks, and have an uninterrupted good draft. The receivers, syphons, and lower coppers, should uniformly extend from the abutment of the chimney, gradually lessening in size to the tache. One receiver containing 270 gallons, will

be sufficient. Two syphons holding the
same quantity each. The grand copper
must be equal to the syphon, that is, to
contain 270 gallons. The second copper
should contain 190 gallons. The third
copper 110 gallons, and the tache 65 or 70
gallons. I will here venture to assert, that
it would be always better to have a second
tache, to be hung to a separate fire, and to
be boiled and worked occasionally by coal,
in case a want of strong fuel might arise,
on purpose to spare the necessity and expedient of bringing home field trash or
brush, than which nothing wastes labour
more, or is more injurious to the cane field,
at the same time making indifferent sugar,
with tedious unsatisfactory labour. I should
prefer shell coppers for manufacturing
sugar, to those whose bottoms are rivetted
to their tops, because they are with more
ease kept clean, and safely scoured, than
those coppers which are rivetted together;
the heads of which rivets, are frequently
burned off by the action of the fire in the

furnace, and then the seams open, begin to leak, waste the liquor, endanger the boiling-house by fire, and the coppersmiths and masons must be called in to put them in order; all which causes vexatious delays, and perhaps loss of materials. A strong high railway should be made, to reach across, between the grand copper and syphons, so that the negroes can pass, to draw the liquor from the syphon cocks with safety. Many horrid accidents befal them, by falling into the grand boiling copper. The parapet wall of the lower coppers should be so high, that the people in leaning to skim the vessels, may not be thrown off their balance, and their hands and arms get dipped in the boiling liquor. Every copper should be provided with a well-cleaned ladle and skimmer, so that no delay may arise in borrowing a skimmer or ladle, to clear the liquor, or throw it from copper to copper, to replenish them, and prevent their burning. The leads of the coppers should always be kept clean, and there

should be a couple of well-washed mops
in the boiling-house for that purpose.
There should be in every boiling-house
four strainers cleanly washed by water. I
should prefer square ones, as they are
more easily managed and fitted to a frame;
one of which, of rather coarse texture,
should be placed and made use of in the
syphons, in drawing the liquor from the receiver. Another should be placed between
the grand and second copper, of rather
finer quality. A third between the second
and third copper; and the fourth between
the tache and the coolers, of a still finer
texture, which should be fitted to the bowl
of the skipping gutter. Every estate capable of making 200 hogsheads of sugar
annually, or from 15 to 18 hogsheads per
week, should have three sugar-coolers,
fitted up in the boiling-house. They must
be made of the best seasoned hard wood
plank, closely joined together, and upon
substantial sills, engrafted in well-cemented
mason work. A receiver or vessel should

be sunk underneath them, to contain any ousings of molasses, &c., that may come from them, and branching gutterings from the lower part, where the sills are placed, should lead those drainings to the receiver. A footway three feet broad, and raised four or five inches above the level of the boiling-house floor, should be made and terraced, for the people who work in the boiling-house to stand upon, close to the parapet of the coppers, to prevent their feet from being tormented by the heat of the floor contiguous to the furnace. The stoke-hole, or place appointed for the negroes to make fire under the coppers, should be spacious ; capable of holding as much trash or fuel, as will boil two skips of sugar, and it should be covered in by a high brick archway, open in front, and joined to the boiling-house wall, to prevent accidents by fire. This archway should be supported by mason work pillars, and continued as far, as the length or range of the furnace, and terminating at the cooling gate. No rub-

bish should be kept in it, or any quantity
of ashes allowed to accumulate, either
under the coppers, or ash-pit. The former
should be kept particularly clean, that
there may be no impediment to the draft,
from the flues, to the syphons and chimney.
Neither should any water or liquid be permitted to be thrown under the coppers, to
damp the fire, (the cooling gate at the extremity of the furnace being adequate to
effect it,) as the explosion arising, from the
two opposite elements of fire and water,
causes such a concussion, as to endanger
the coppers in their places, and shake the
perhaps too crazy mason work about them.
Instead of two hanging copper lamps,
which are made use of in the boiling-house
at night, close to the lower coppers, and
the heads of the people there, to furnish
them with light, I prefer a globe lamp,
with three good burners in it, well supplied
with oil. It should be hung in the centre
of the boiling-house, at a height to prevent
its being broken, and sufficiently low to

diffuse general good light. This will serve the purpose of giving light to the man on the syphons, as well as to those at the lower coppers. It will likewise cause a saving of oil and wick, the consumption of which will be one-third less, the light better and more lasting, and prevent the thieving of the negroes, who watch every opportunity, not only to steal the oil, but the wick soaked in it. One pint of oil will be enough for the globe burners every night, whereas it takes near a quart every night, when the boiling-house is at work, to supply the lamps for the low coppers and syphons. I would recommend exemplary cleanliness in the boiling-house, and that the walls be kept well white-washed.

The curing-house, a building which should ever be attached to the boiling-house, as its receiver-general, comes next to be taken into consideration, as to its consequence and utility. This structure should be strong, durable, and built of the best mason-work, timbers, plank, &c.

It should be so constructed, as to be placed between the boiling and distilling-houses, receiving from the one, and giving to the other. Its platforms should be elevated three feet above the level of its own, and boiling-house floor, which platform should be composed of the best seasoned stuff, the rangers of strong scantling, without knots or blemish, and distant from each other about sixteen inches, and well levelled, with uprights to support their centre. The underneath inclining plane, which catches the molasses, as it runs from the sugar hogsheads, should be laid transversely to the rangers, and should be made of season-ed inch boards, the lower board to be laid down first, with a gentle inclination to the adjoining guttering which gathers the mo-lasses. The lower edge of the next board must be laid about an inch over the first, likewise with a gentle inclination, and so on, each board to the height of the plat-form. The gutterings to receive the mo-lasses from the platform, should incline to a common centre, and perpendicularly underneath such centre. The molasses cistern or receiver should be placed, of sufficient magnitude to contain some hundred gallons of liquid. I prefer a long wide cistern, well put together, with bracing bars and keys, so that the molasses can with little difficulty be drawn, or carried to the distilling-house. A curing-house should have several glazed windows on each side, to rise and fall occasionally, to admit air in dry weather, and to exclude it in rainy or damp weather. The windows should be secured with good stops inside, and a strong wire lattice to cover them on the outside. Care should be taken, neither to have these windows open in damp weather, or at night, for nothing destroys the sparkling grain of sugar more, than dampness. It even changes its colour from a bright straw, to a dingy, sandy hue, and makes the sugar sink unnaturally in the hogshead, which must necessarily be filled up before sent to the wharf. This is a vexatious oc-

currence to an overseer or proprietor, who
calculated upon a certain number of permanent casks, or hogsheads of sugar. Good
light sail-cloth coverings, painted on the
outside, should be in every curing-house,
to draw over the cured sugar. If accidents
from fire were not so much to be dreaded,
and guarded against, on every sugar estate
the acquisition of a fire-place, situated cen-
trally in a curing-house, for a fire to be
made in occasionally in damp weather,
might greatly contribute to preserve the
grain, quality, and quantity of the sugar,
that, at times, unavoidably must remain for
months in the curing house of an estate,
before it is in order to be sent to the wharf
or shipped. Every curing house should be
kept perfectly clean, and often white-washed
inside with strong lime, to prevent the in-
road and propagating of cock-roaches (a
creature of the beedle tribe) who eat and
mar the sugar.

The distilling-house on a sugar estate,
should be situated at the extremity of the

curing-house, or in a mean distance between it and the boiling-house, and lying
a little way from that end of the boiling-
house, where the sugar-coolers are placed.
All three houses may be joined together,
embodied in a building, if the havoc occasioned by fire was not to be contemplated, and the dreaded evil to be guarded
against. This building should be extensive,
and well made of the best timbers and
mason work, with a tank inside the building, capable of containing from five to ten
thousand gallons of water, for the still
worms to be immersed in. This tank should
have a constant influx of cold water, and
an outlet to carry off the same quantity;
a pipe should be introduced into the distilling-house, not only to supply this tank
with water, but to wash the vessels and
mix the liquor with, when wanted. This
pipe should not run all round the walls, as
in some distilling-houses is the case, as
thereby, (having cracks or holes in it) to rot
and moulder the walls, &c. but be intro-

duced through the wall plate, and run parallel with the centre of the tank, in a horizontal direction ; the pipe should perpendicularly dip into the tank to the bottom of it. A branch pipe should extend from that to the mixing cistern, and another to the main guttering that supplies the liquor still, both with stop-cocks attached to them. I have ever found that cisterns sunk in stiff mason work, and rammed all round with good clay, free from stones, to within two or three inches of their top, gave not only the greatest returns of rum, but fermented best, and that there was little loss sustained by leakage. Vats or keyed cisterns fixed on sills, are more liable to leak in the dry months, do not retain fermenting heat as well as the sunk cistern, want constant repairs, and are not so useful to the workmen. The inside of a distilling-house, in my opinion, should be in three compartments ; one for the fermenting cisterns, one for the tank, (and where the worm-cocks give vent to the low

wines, and rum from the stills, likewise
where the molasses and mixing cisterns
should be placed,) and one for a small rum
store, to hold three butts, that would contain from twenty to thirty puncheons of
rum. The mason work for the fermenting
cisterns, being raised from about the centre
of the house, carried its entire length, and
the cisterns being ready to be put down,
each of which should contain the same
quantity of liquor as the low wine still, they
should be separated about fourteen inches
asunder, and nicely levelled ; so that their
frames, when the work is finished, will appear on an even surface with the top of the
mason work. The intervals between each
should be neatly terraced to their surface,
and the whole made to appear a neat, compact, firm work. The distilling-house being
in a manner cut in two, by this mason wall
inclosing the fermenting cisterns, the other
half of the house comes next to be disposed of and laid out. The tank necessarily takes up a great part of it. There

should be a space of three feet left between it, and the wall inclosing the fermenting cisterns, for the low wine and rum cans to be placed under the worm-cocks, and give ample room for a man to pass and repass, to carry them to and fro. At one extreme end of the tank, that next to the boiling-house, there should be a cistern to hold redundant molasses, that the molasses may be drawn from the curing-house, by a gutter or pipe, to be convenient for mixing liquor; and likewise, a mixing cistern should be placed there, the molasses cistern below, and the mixing cistern above it. At the other extreme end of the tank, the low wine butt should be placed, close in with the wall that separates the rum still from the inside of the distilling-house, and parallel with that still. It should be fixed on a firm still, elevated enough, to give a fall for the liquor to the still, and to preserve its bottom from any foulness that may rot it. The remaining part of this end, or division of the house, between the low wine butt and end wall, should be taken up with a small rum store, separated by a partition, with a door and lock, from the low wine butt and other parts of the house. As theft is often practised by the negroes, on this tempting liquid, they should be prevented from so doing by all possible care. This rum store should have two or three good sized butts in it, to contain at least twenty puncheons of rum, and a strong door, opening to the yard, from which the waggons are loaded, when rum is ordered to be sent to the wharf. This door should have a good lock and bolts, and in the centre of the door a small imbedded frame, with a pane of glass to admit light, and the inside secured by two or three cross bars of iron. All the butts should have a coat of thick paint once a year, to prevent worms from eating through them. Strong white-wash will answer. Two cisterns to hold skimmings should be made, (one larger than the other,) outside the distilling-house, opposite the boiling-house, which may be covered over with a good shed, and boarded in, with a door and lock, to prevent theft by the negroes or hogs ; the skimmings may be drawn by a pump, from the lower to the top cistern, and then left to subside for some hours to clarify itself, before it is drawn off to the mixing cistern. The skimmings may be drawn by a pipe with a stop-cock to the mixing cistern, and a branch-pipe from the skimming pump, be introduced through the wall, into the molasses cistern, within the distilling-house, reaching to the mixing cistern, with a cock to stop the working of the sucker upon the skimmings, while the molasses is drawing, and the same with the molasses when the skimmings are drawing.

The distilling-house, stoke-hole, or place where the stills are hung, should have a lofty shed over it, supported by masonry pillars, at one end of which, the rum still should be set, parallel to the low wine butt. At the other end, next to the boiling-house, and parallel to the guttering, leading from

P

the fermenting cisterns, the low wine still should be placed, leaving a space between the still and the end of the shed, for a dunder cistern, which should be sunk in the ground, and made large enough to contain, more than double the quantity the still does. Another smaller dunder cistern, should be built over the bottom one, (to contain as much as the mixing cistern) in order to draw the dunder by a pump, from the bottom to the top cistern, where the dunder should remain for some hours to cool and clarify. The top dunder cistern should be a little higher than the mixing cistern, to give an inclination for the liquor to run, which may be drawn by a pipe, introduced through the wall, and a stop-cock affixed to it, to the mixing cistern. The liquor from the mixing cistern may be drawn off by a short pipe, with a stop-cock to it, into a temporary gutter, that should lead to the fermenting cisterns. The mixing cistern should always hold as much liquor, as will fill two fermenting cisterns.

The stills on a sugar estate are generally
of a large size, the low wine still containing, from a thousand to two thousand gallons, and the rum still, from five hundred
to a thousand gallons. Being of such large
dimensions, they consume a great quantity
of fuel, mostly large heavy wood, which
requires great labour to hew it; and the
carriage is distant and heavy. Shallow broad stills would be the best to be
sent out, and fixed in every estate, because
the fire soon takes effect on the liquor
spread within a shallow still. And when
the ebullition takes place, it requires little
addition of fuel to keep the distilling liquor
running through the worms. High broad
goose necks answer equally as well as large
still heads, and are more portable, handy,
and less expensive. The goose necks should
be soldered to the top of the still, and a
hole cut in the top of the still, (with a well-
fitted copper cover to it,) in order to admit
a negro going down to scour the still.
Round the hole, a firm strong copper rim

should be soldered, an inch and a half deep ; and the cover, should likewise have a firm, strong copper rim, soldered to it, to fit neat and close within the rim of the hole. This cover should have two strong handles, to take it off and put it on with. When fixed down, it should be fastened with an iron bar, and locked to a staple, then covered over with clay, till the still is worked off. A temporary gutter should always be ready, to throw water into the stills when worked off and emptied, and to clean them out. The main guttering inside the distilling-house, should be ready filled with liquor, or the low wine ready to descend into the stills, to load them without delay ; stills should never be quite filled ; six or eight inches should be left empty at the top, to prevent accidents by blowing, or sending the liquor or low wine down the worms, instead of low wine from the liquor still, and rum from the low wine still.

The worms, if possible, should go round

the tank, and converge (when they make
a circuit of it) to their respective goose
necks, and then dip with many coils to the
bottom of the tank.   By having the worms
large, of good length, and making many
revolutions in the tank, with a good supply
of cool water descending to the bottom
of it, the low wine and rum will distil
cool and clean.   An outlet from the top
of the tank should always be made to
carry off the warm water, which rises to
the top, which is occasioned by the heat of
the worms.   A door with a lock to it,
should be made to the distilling-house wall,
to pass from the tank to the top of the
stills, so that the young man, who superintends the distilling-house department, may
occasionally go to view what is doing in
the stoke-hole, and that the fire-maker is
performing his duty.   A moving light
pump, of sufficient length for the fermenting cisterns to draw the liquor to load
the stills, is very requisite, and a skimmer

and ladle to clean the liquor with. Cleanliness is the life and soul of a distilling-house, for good rum, or good returns, cannot be produced without its purifying virtues.

CHAP. V.

The sugar-cane, in its primitive state,
seems to court the favour and auspices of
the great luminary of day. In its infancy
it is cherished by it, in its youth it is
invigorated, and in its progress to maturity
it is ripened and supported. So many
favours proceeding from this great enlivener of nature, induces the sugar-cane
to seek, as it were, a situation to benefit
by its sustaining powers. Providence has
so diversified the landscape, that a variety
of grounds present themselves to the view,
of hill and dale, promising welfare to this
benevolent, useful, and desirable plant, and
to attract and induce (by having a genial,
open aspect) the planter to till such a piece
of land for the culture of it. I must, there-

fore, beg leave here to premise, that it is an essential point in plantership to study and choose ground which may possess an aspect, and obtain the influence of the sun. The plants come up sooner and better by this means; swampy, cold places are dried up quicker; the dormant virtues of the land are called into action by its generative warmth; the sugar-cane appears strong and healthy, shewing a luxuriant promise, and the juices are richer, and with more speed brought to maturity, and make better, stronger-grained sugar. So impartial is this luminary in giving and distributing its nourishment to the vegetable creation, that some time of the day its rays will strike progressively upon most parts of a cane piece, save such places as are sunk in gullys or dells, nature excluding them from its bright beams, yet giving and granting them the virtues of the air that is heated and rarified by it. The bounty of the morning sun is mostly to be prized, mostly to be wished for on cane-pieces. It dissipates the rigours of a cold, damp, bleak night, makes vegetation glad, spontaneous, and thriving; and generally when it shows upon land early in the morning, its visits are prolonged till evening. Those spots mostly exposed to its influence, that happen to be gully or poor, I would invigorate and enrich with solid manure, that every advantage may be drawn from it. Not that by so doing, other parts of the cane-piece should suffer neglect, for the whole should have impartial justice done to it. Even the gullies where canes are most apt to lodge, are generally out of sight; and where the canes degenerate to suckers, and seldom produce rich juices, or good sugars, should have their share of care and attention. The canes in those places should be well trashed, or bolstered with trash, to keep them from lodging, and hunted for rats frequently. Having now urged the necessity and the advantage of an open aspect for cane cultivation, I shall proceed to point out what kind of land I presume to think best adapted to produce good canes and good sugar. Though we cannot always have the choice of ground, a great variety of soil presents itself to try the skill of the planter. Ground nearly level, having a gentle inclination to drain it, and pass off superfluous water, is best for a cane-piece, as the carriage to and from, and manuring of it is easily effected. A thick stratum (from twelve to eighteen inches deep) of dark brown, friable, unctuous soil, upon a yellow, moist, though warm clay, which may easily be turned up, is the best to produce good, strong, long-jointed sugar-canes, and bright-coloured, hard-grained sugar. Such ground often presents itself to the planter; yet from the exhausting nature of the sugar-cane, its appetite, and great ability in sucking the strength of the ground, for the support of its large, fibrous, prolific root, and tall, juicy, numerous stems, it behoves the provident planter to keep up the stamina of the soil in those rich lands, by manuring every time the piece requires replanting. It would even

13

be well to manure the ratoon, to prevent few recurrences of turning the land up to replant it, for it weakens the land, if it is ever so good, to turn it up and break it for a fresh nursery of plants. The ratoon, by such management, and good care, may make equal returns to the plant-cane, and far better sugar; with less boiling, labour, and fuel. The parishes of St. Thomas in the East, St. Andrew's, St. John's, St. Dorothy's, St. Mary's, Trelawny, St. James's, and Clarendon, in Jamaica, possess such a soil in many places. So luxuriant are the canes thrown up from this soil, that they often tempt the planter to give them extra trashing, which should never be done in dry weather parishes, or districts; as thereby the cane is dried up, burned, and shrivelled. The trash often falls off of itself by excessive heat, when by its dryness, it loses its adhesion to the cane joints, nature thus relieving itself. And this is sufficient, after the first hand trashing, in dry parishes. But in moist parishes, such as Portland,

St. George's, St. Mary's, St. Thomas in the East, &c., where canes grow from such a luxuriant, heavy, strong soil, are very thick, long-jointed, tall, and top-heavy from their weight, three trashings will not be too much to give them after the banks are levelled, for the influence of the sun then becomes necessary and apparent, by ripening and thickening the juices. It is one of the most glorious sights in the vegetable world to behold a ten acre field of plant sugar-canes, rising from such a soil, which has justice done to them by the planter, in one of those temperate parishes, after it has received its last trashing, and within a couple of weeks of being declared fit for the bill. Nothing in the world of vegetable substance, that has tried the art of man in raising, has no noble, so generous, and luxuriant an appearance.

Upon level grounds where the cane-pieces are easily laid out, roads formed, manure carried, and the crop brought to the mill, I would not have too much room taken up with intervals, and only those formed, that may be necessary to give access to the farther cane-pieces. Extra or superfluous intervals waste much good ground, which might be planted with canes, and add no beauty or safety to the cultivation. A general interval of twenty feet broad, between two cane-pieces, is quite sufficient; cross intervals are of little use, because those general intervals partition the cane-pieces from each other, mark their limits, and are commodious enough for waggons to pass to and fro in carrying canes, &c. A gang of negroes when set in to cut a cane-piece, must always make an opening for cane carriage, as they cut down the canes. But I have known many overseers intersect a level cane-piece with several intervals, losing by such unmeaning tracks, one eighth of the cane-piece, or ground that ought to be in canes. Neither would I waste any ground, between a cane-piece and the surrounding fences, farther than would be necessary to allow a negro to clean the fences, and bank up the head rows. In my opinion, the lining out and levelling of water trenches, to carry off redundant moisture from a level piece of ground, should be much attended to, and performed by a skilful hand. So should the lining out the ground there for cane-holes be done with judgment, that the holes may all be of an equal breadth, perfectly straight, and no cross or half rows intervening to spoil uniformity, or break the formation or working of an effective field gang. Whatever manure is necessary to be expended upon a rich piece of level land, should be carried to the ground, and put up in heaps, at proper distances from each other, and covered with trash till wanted, when the ground is to be planted with canes, and before the piece is lined or trenched. Nature, in a great measure, points out the course a trench will take : however, she must be assisted by art sometimes ; and if impediments of no great magnitude arise, they should be cleared away, or a circuitous track levelled for the trench. In order to avoid and obviate such a laborious

service, I would cut no more trenches in
a level, or a cane-piece, than what are absolutely necessary to keep the land free
from water, or give proper vent to a spring.
It has been an old custom with some over-
skilful planters to tire their invention, by
traversing a cane-piece, both length and
crossways, with a multiplicity of useless
trenches, as if to shew or display nice
plantership by this intersecting game, or to
stamp upon the vulgar novice an high
opinion of their talents. By such prodigality, they bring poverty on the land they
are commissioned to support and maintain.
Nothing impoverishes land more than useless trenches, or a great number of them;
nothing is more likely to break it into deep
inequalities, which destroys a fine surface;
nothing entails sterility on the soil more,
carrying away, by every discharge of heavy
rain, the best of the soil through those
superfluous drains, and leaves to view
scarcely any thing but a forlorn, hard,
cold clay, which, by its water-washed, white,

sickly appearance, seems to bewail the loss of its warm productive powers. Survey then minutely the ground before it is trenched, and let none be cut, but what are expedient to keep the land clear from cold, swamps, or springy overflowings ; using as it is lined out a level for so doing. After the trenches are dug, the piece should be cleared and stocked up of cane-roots and grass, &c., previous to lining it for cane-holes. A level piece of rich land should never be lined into cane-holes of less than four feet and a half, or five feet wide, and a large wooden square, and flagged line made use of, to strike and measure off the rows and distances, setting off from the heading of a straight interval. Nor should the ground be turned up deeper than the surface of the clay, or within an inch of it ; because the clay being cold, or perhaps sour, may either retard the growth and sprouting of the plant top, or burn and wither it in the ground. The plant cane-tops should be carried, and put down in heaps,

along the interval, the negroes should strip
and trim them there, and bring them to the
spot they are to be planted in, that the banks
may not be broken down by mules, or the
cane-holes broken up by them. The weeding gang should attend with dung-baskets,
to carry and drop the manure into the
holes. The great gang having stripped,
trimmed, and carried as many tops as will
plant out a breadth, they should begin to
clear the bottom of the cane-holes well
with their hoes, weed the banks clean, (the
weeds to be put on the top of the banks,
with their roots turned up to the sun, in
order to kill them,) haul the loose mold to
the banks, and the manure being dropt
before them in the cane-holes, they should
spread it pretty thick in the bottom of the
holes. Care should be taken that no tops
are brought to be planted but such as are
fresh, have good eyes, have not been
arrowed or sackered; and the longer the
joints are, the better. I would leave little
or none of the soft, furry part of the top

to the plant, as a sucker may proceed from it, by which the nourishment is extracted, and extorted from the sprout that grows from the eye. All these preparations being ready, and the negroes formed in a straight breadth, a negro to each cane-hole, set them in to plant the tops. First, the manure being spread in the holes, let each negro cast a little light mold over it, that the plant top may not be scalded, and to keep the sun from absorbing the strength of the manure. Then let each negro take two clean fresh tops, and lay them firmly in the bottom of the hole, across it, on the soiled manure, and about eight inches asunder, the bottom or hard part of one top opposite to the soft, furry part of the other, then cover over those two plants lightly with mold, suffering no weeds or grass to remain on them, as weeds shake well from any earth; and place them on the top of the banks, with their roots exposed to the sun, to kill and wither them. When those two tops are thus planted, and each negro is provided with a measure-stick of fourteen inches long, make each of them measure a space with it, from the farther planted top, and put down two more good plants on the soiled manure, as the former were done, covering them with mold, and weeding them clean as the two former plants were managed; and so on progressively to plant out the breadth till it is finished and completed. Care should be taken to drop no more manure in the cane-holes than what will suffice for one day's planting of tops, as the sun may injure it by being exposed to it. The trenches and intervals should be cleaned of weeds, as the gang proceeds in planting, allowing no grass to grow in the interval within six feet of the canes. When the first breadth is thus planted out, let the great gang strip, clean, and carry as many good tops as will plant out another breadth, and so on till the whole of the cane-piece is planted with cane-tops, leaving every thing clean and secure upon it, for Providence in its goodness to preside over.

The fences round this and the adjacent cane-pieces should be cleaned, closely made up, every gap stopped, and the gates well hung, with iron fasteners to open with occasionally. The planting of corn or French beans on the banks of such a piece, at five feet asunder, can be no injury to the canes. The crop of corn takes no additional trouble to raise, except the planting and breaking it in, a work of little labour, and amply paying for such trifling expence by the plenty which ensues from it. This crop of corn is taken off in four months from the time of planting, the same period that the plant-canes require the banks to be levelled on, or round them, as the last molding they are to have till ripe, and cut down to make sugar. In rainy, moist weather, the young plant-canes require almost constant weeding, so quick and excessive is the growth of sour grass and weeds about them. On no account should they be neglected. As soon as weeds or grass appear growing among them, the

weeding-gang should be set in with small
sharp hoes, to clean them of such foul
annoyances, using great caution to pull
out, with their hands, the weeds that intrude between the canes, and placing the
weeds on the banks, to be withered by the
sun, first shaking their roots of any earth
that may be about them. The banks should
be well cleaned of the growing weeds, and
a little mold drawn from them round the
young plants, according to their growth, to
impart gradual nutriment to them. Some
supplies may be wanted ; I think but few
for the first cleaning, as the tops are not long
in the ground ; but at the second cleaning,
which may be requisite shortly after, I
would set in the second gang, (a more
sensible people than the weeding gang,) to
go through them ; who should search well
for any dead tops, and replace them by
lively fresh tops, molding such supply tops
lightly, and putting a little dung underneath them. This gang should bring out
dung-baskets ; dung should be ready for

them on the interval, and as they proceed
in cleaning and supplying the piece, they
should throw in some of the manure through
the cane sprouts, and draw mold round them.
No supplying should be after this, except a
great drought or blast comes on, which
Providence sometimes orders; then the
persevering ardour of the planter comes to
be tried, and replanting the piece may be
its fate.   Even continual rain for a length
of time may strike a chill into the ground,
and retard its animating functions so long,
that the tops may rot.   I will venture, however, to assert, that after such a cane-piece
has had its second cleaning, manuring, and
molding, it will soon begin to shew abundance of vigorous, strong sprouts, of a deep
green colour.   Their growth will be quick,
waving abroad, with saw-edged leaves, and
their young stems will be stiff to the hand.
The third cleaning and molding they receive,
the second gang should likewise perform.
Then the plants will exhibit some white
small tender joints, from which may pro-

ceed wiry, small, dry particles, or leaves of
trash, which should be gently pulled away,
if free to drop, and one half of the bank,
mashed fine with the hoe, should be given
and drawn round them ; taking care to free
them from all interloping weeds and grass
which may spring up, and placing them
(after shaking the earth from their roots)
upon the banks, with their roots turned up
to the sun, to wither and kill them. Few
weeds, after the third cleaning, will grow
up to annoy them ; they nearly bid defiance
to their attacks. They spread with amazing
luxuriance over the surface of the ground,
claiming the traveller's attention as he
passes by. The fourth cleaning and moulding finishes the operation, for they then
must draw the veterans of the estate to
behold their rising beauties, and minister to
their prosperity. The great gang should
be set in at this period, to level their
mother earth, the banks about them, cutting up every thing with the hoe by the
root that may in any measure blemish their

lovely appearance, or retard the quickness
of their growth. Any loose trash that may
be on them, should be pulled away with the
hand. This will give them freedom, but
nothing to injure their joints, or expose
their still tender green stems to the parching sun, or the asperity of the weather.
The powerful hand and hoe of the great
gang should be brought into action, to
chop the remnant of the bank fine, and
haul the mold about them, but not to
raise any hills to their joints. This some
old planters do, in the vain hope of increasing their growth, or prevent them
from lodging; not considering that by thus
burying their joints, they cause them to
throw out suckers, thus checking their
growth, vitiating their juices, and making
them slender and spindling. A piece of
ground of this description, planted in the
before-mentioned way, the latter end of
April with cane-tops, will have the May
season to bring them up, and escape
arrowing in the October following, or the

arrowing months, and attain their full growth and age, without being checked or any way empoverished, by this absorbent, periodical malady of the sugar-cane. As I said before, in moist parishes, with temperate weather, such a cane-piece, after the banks are levelled on it, will bear three trashings before it is fit for the bill, or to be cut down to make sugar of. But in no instance, when they are ordered to be trashed, should any of their joints be bled, (a common phrase with planters, when green trash is forced from the joints, and their fibrous veins exposed to the sun, and made to drip their vital juices) or any but the dry, drooping trash pulled away. The joints which lie under the dry trash, declare their own hardihood, and being cleared of it, the sun and air will gradually swell and ripen them. The aspiring sugar-cane will sometimes require proping, to keep its tops from coming to the ground, so heavy do they become, from the fattening care that has been taken of them. The dry trash which has been taken from them, should in such cases be made as a pillow for them to lie upon ; some should be put near their roots, and some in the middle of the rows, to prevent their curved tops and stems from lying on the bare ground, where they might be induced to adhere and take root. Thus, after receiving the third trashing, in about two weeks, when thirteen and fourteen months old, they become sugar-canes in the true sense of the word ; their roots declare their inability any longer to support their stems, (as they are pregnant with embryo ratoons,) they require the relief of the amputating bill. The overseer must obey this natural summons, and direct them to be cut down, carried to the mill, and made sugar of. A piece of such ground, manured, planted, and reared in sugar-canes in the before-mentioned manner, will often give from four to five hogsheads of sugar per acre. It will yield nearly as much in the first ratoon, and preserve its stocks for subsequent ratoons for

five or six crops, thereby saving the expense, labour, and trouble of digging the
land often to replant it. It will empoverish
it by so doing, and spread the cane cultivation over an enormous space of ground,
forcing an indifferent crop from it; causing
twice as much labour as is requisite, over-
working the cattle and mules, and presenting a large field of canes to the sight,
which gives but poor returns. I have thus
presumed to point out the utility and happy
results arising from manuring even rich
lands well in Jamaica, for cultivating sugar-
canes, and the still happier termination of
the overseer's care, by planting it with
good tops in the proper spring season; and
taking special care of the plants when they
come up, till they are fit to cut, and carry
to the mill. I shall next proceed to point
out the plan of manuring distant, hilly,
poor cane pieces, so as to save the necessity
of too much ground being taken up to obtain a crop.

It should ever be a maxim with a plan-

ter, who has the management of a sugar
estate, to bring his cane cultivation into as
close an order, and as contiguous to the
works, as the nature of things, and the
required crops from the estate, will admit
of. Straggling cane-pieces have a confused
appearance, exemplifying a want of connection or method, and principally evincing, either the inability of the overseer,
in point of professional skill in using manure, his vanity in having a large, unproductive cane-field, or his want of cattle,
mules, and materials to produce it. In a
former part of this work, I alluded to the
great necessity which exists of having an
adequate number of working and breeding
stock on an estate. For not only there is
less mortality among the stock, by such a
policy, having frequent spells to relieve
them, but a combination of circumstances
occurs, making a sufficiency of manure, to
return, and retain to the soil its strength
and virtue. Labour is hereby saved by not
digging too much, and expence is prevented

by not resorting to jobbing. The negro strength of the estate is thrown upon the plants and ratoons, to keep them clean and do them justice. The land yields one-third more, by the acre of sugar, and the sugar of a better quality. The canes are faster, and with more ease, brought to the mill, take less boiling, and less fuel. The field trash is not obliged to be taken off the cane-pieces, to furnish a supply of transient fuel, by which the poverty of those grounds is increased, and labour lost in the carriage. And, lastly, more negro labour can be spared to clean the pastures, (which sometimes run to ruin for want of it) and procure copper-wood, build lime kilns, make roads, &c. All this can be done by making plenty of manure, and confining the cane cultivation to a moderate sphere. In the course of this work, I shall beg leave to relate one or two circumstances, in which the resident agent had a principal hand in overrunning an entire estate with numerous cane-pieces, extending an im-

mense field of canes, cutting down, burning, and destroying valuable wood land, &c.
for the sake of stretching out a large field
of canes, which, in the end, did not produce more, or even so much, as when one-
third less land was in canes, and the jobbing
six times as much. I would begin, by having
cut and brought home from the woods, as
many mortice posts and rails as will make
three temporary or moving cattle pens, to
contain from eighty, to one hundred head
of cattle with ease; so that they could lie
down, get up, move about, and feed upon
what provender there might be in the pen,
without incommoding one another. I shall
premise here, that whatever manure may
be made about the works, from the mule-
stable and cattle-pen, should be heaped up
in pits sunk there, or contiguous to it, of
eighteen inches deep, that the urine, or
strong material of the juice that issues from
the dung, should be retained, and not
suffered to run off, (which is usually the
case,) losing by such neglect or oversight

16

the most essential and best part of this invaluable article in husbandry. This dung made at the works, and so heaped up in sunk pits, should, after the heap rises to four feet, be chopped with sharp hoes by a gang of negroes, moved to another convenient pit, and the moisture or juice that run through it be thrown on, and mixed up with it; then a layer of strong, rich earth, of eight inches deep, should be thrown and spread on the top of it; after which, and over the earth, more dung from the cattle or mule pen may be put on it, for three or four feet higher; then a second time chopped, and removed either to the former or another pit, taking care, whatever juice may be at the bottom of the pit, to have thrown on, and mixed through it. After this second chopping and removal, the heap should be topped with another layer of earth of eight inches deep, and covered over with trash or cane bands. All the moisture being retained in this body of manure, it will soon vehemently

ferment, intestinely and externally rot, and
become a solid heap of well-digested, available, and profitable manure. It will be
necessary in showery weather to take the
covering of trash off it, that the rain may
penetrate through, and assist the fermentation and putrifying of it; but in dry,
parching weatner, this covering of trash
should be replaced; and it should be the
special concern of every overseer on a
sugar estate that not the smallest particle
of any thing whatever that can contribute
to make or increase manure should be
wasted or neglected, either about the works
or elsewhere. Every convertible substance
for that purpose should be scrupulously and
carefully thrown into the manure-pits. In
this manner should manure be made and
accumulated about and at the works of
every sugar estate. This manure so made
at the works, should be expended on the
near cane-pieces and bottoms, and generally will be sufficient for that purpose,
except here and there a gully spot, which

will plead for the aid of the cattle-pen, to
send its penetrating juices into its meagre
earth. This call of nature should always
be attended to, having the posts and rails
at home, with what apparatus may be wanted
of ties and stockadoes to complete it. It
should be considered whether the land is to
be manured for a spring or a fall plant, as
the planting seasons are divided into spring
and fall, the first beginning, as it were,
in March, and ending in June or July, the
latter commencing in September, and ending in December. The first having the April
and May seasons, (or showery weather,) to
promote and bring up the cane-plants, the
latter the October rains to swell the earth,
after the parching months of August and
September, which proves generally effectual,
with good manure, to bring up strong plants.
I would leave the distant, hilly pieces for the
spring, the near hilly pieces for the fall,
and the bottoms for the spring or fall plant,
agreeable to the nature and situation of the
ground. The spring plant comes first to

R

be considered, and the number of acres which should put in, to provide for a regular good crop the ensuing year. I would begin to manure, by cattle penning with those temporary pens for the spring plant, in the months of September and October, and set up two cattle-pens on the piece I intended first to turn up and plant, by lining the ground out for them. The mortice-posts should be seven feet asunder; pitching the pens on that part of the ground which proved to be the least productive, or where the soil was light or cold. On those parts I would let the pens remain for a fortnight or three weeks, all which time I would be particularly provident in supplying them with plenty of green food, either of long cane tops, guinea grass, or both mixed, placed all round the rails, inside of the pen, for the cattle to feed on; and have the bottom of the pen strewed with dry cane trash, old guinea grass, the choppings of intervals, &c. for the stock to lie, dung, or urine on; in order to ac-

cumulate by such means a considerable heap
on the spot at the termination of three weeks.
In each of those pens I would enclose, every
evening, the cattle, (except those which
may be intended for road work,) the breeding and young cattle in one, and the working cattle in the other, (the road cattle to be
penned at the works,) so that no disagree-
ment may take place among them, and they
may have plenty of feeding. In the morning they should be dressed and reckoned
there, moving them about for a quarter of
an hour, to give them an inclination to
dung and urine in the bottom of the pens
before they are turned out to graze. By
thus confining the cattle in these temporary
pens at night, with plenty for them to eat,
strewing the pens with trash, withered
grass, or any slips, rubbish, and weeds, that
can be procured by invalid negroes, a heap
of manure will be raised, not only for the
spot where the pen is, but to throw on
other parts of the piece where the pens
are not placed. After three weeks' time

the pens should be moved to other poor spots, closing up the mortice-post holes where they stood. Then the second gang should set in, and cover over this heap of manure with good earth, which may be found in some gully or waste spot contiguous to the piece, at least six inches deep over the surface of it. This heap remaining closed up with earth, confines the moisture within the body of the manure, any showers of rain that descend penetrate through the earth into it, and cause an intense fermentation, which soon corrupts and rots all the different substances it is composed of. In about a month or six weeks after the first pens are removed to other parts of the piece, I would set in the great gang with sharp hoes to chop and mix up this stuff, which has been made in the first pens ; which being done, they should heap it up high, make it lie compact, and then cover it with trash to prevent the sun from injuring it. Sometimes it may require a second chopping to make it unite and become more unctuous, which is discretional, and easily performed. The pens should be removed every three weeks to different spots, well fed, supplied, and covered with quantities of vegetable substances that will soon ferment, rot, and unite themselves, by being closely condensed together. In the course of eight or ten weeks a piece of ten acres of poor land will have a sufficiency of manure ready made upon it to answer the purposes of planting with ability, in the ensuing month of March or April; which is early enough to put in a seasonable spring plant. The manure made on this piece will be rotten fat stuff in the following January or February. I here beg leave to request my readers to calculate, that this first piece of ten acres, having taken ten weeks to make an adequate quantity of manure upon it, it falls out then that, having began the first of September to pen it, it will be completed by this system, by the middle of November following. Having by this time judiciously

chosen another piece of poor ground to
throw the pens over for the same benefit,
remove those two pens to it, taking care to
stop up all the post-holes on the former
piece, and have the manure made up upon
it; taking the same before-recited method
with the other piece, you will finish the
manuring of it likewise by the middle of
the following February. This is taking it
for granted that it will measure ten acres.
Thus the planter may go on progressively
to another distant hilly piece, without de-
viating from the foregoing rules. He will
then have thirty acres of such land, in three
distinct pieces, ready manured by the middle
of May, in order to turn up for a spring
plant; beginning with the first piece to dig
and plant in March, the second in April,
and the third in May, all coming in, as it
were, in succession. All this is beside what
near pieces may be deemed expedient to be
put in for a spring plant, which may be
manured, in a great measure, from what
dung is made at the works, and carried out

to them. I must here remark, that the very time I propose to begin cattle-penning for a spring plant, is the providential time of the year, which nature, by her temperate, seasonable weather, blesses the planter with abundance of feeding for the cattle, as if by such bounty inviting him to such an undertaking. Guinea grass, common grass, and herbage of every description natural to this climate, have a spring and luxuriance, from October till May, which is astonishingly prolific ; so that no want of green and dry provender will happen if the planter is as provident as nature, in assisting her spontaneous exertions, by keeping his pastures clean. The spring plants, and the ratoons coming ripe at that time, are mostly depended upon, in moist parishes, to make a crop from. The fall plants and ratoons coming ripe early in the year, are looked up to, for a crop in dry weather parishes. Not that I altogether agree with old practitioners in moist parishes, that the fall plant cannot be brought in to answer equally as well as in dry parishes, to make the bulk of the crop; and should be, with equal ardour, followed up and depended upon. Indeed if they would go more on the fall plant in the north side of Jamaica, from Portland to the west end of St. Ann's, than what they do, they would get equally as good returns, by manuring high, as they do from the spring plant. Their cane pieces would become ripe in quick rotation, and they would finish their crops by the end of May or June; have most of their sugar shipped in August, and in the mother country, and sold, before Christmas; whereas, I have known the estates in Portland, St. George's, St. Mary's, and St. Ann's, to have their mills about, (with only intermissions for digging and planting) from January till within a day or two of Christmas; and their cane pieces becoming ripe, with straggling irregularity. The vicissitudes of the weather seem to tell them, that they should adopt the mode of bringing in their crops to be taken off without intermission. To

behold them working the mules, nearly up
to their bellies in mud, carrying distant
canes, from the beginning of October till
December, is distressing. For during this
period in those parishes, the weather is
rainy, and very unfavourable for cane carriage to the works, so that the poor animals
are miserably reduced and cut up. Besides,
a misfortune equally to be dreaded is, that
the canes cut at this season, seldom stock
or ratoon well, the cane-pieces are mauled,
ploughed up, and destroyed, by the mules'
feet ; large pools of water encircle the
cane-roots, which are indented and broken,
by the stock going into them to be loaded;
and immense labour expended, in supplying
those pieces afterwards, which are thus cut
up.   Neither can the sugar be shipped,
till the return of some of the ships from
the mother country, in the month of February following.   All this while, it is lying at
the wharf or in the curing-house, sinking
and perishing in the cask, which must be
filled up or repacked, before it is in a con-

dition to be shipped. This kind of management concludes with comfortless bad
roads, which are rutted up by bad weather,
for the heavy ox to drag the sugar to the
wharf, over a distance of perhaps 10 or 15
miles, making them a way-worn, weather-
beaten, emaciated set of beasts.

Relative to manuring for the fall plant,
by temporary, moveable cattle-pens, I
would without delay commence so doing,
the first or middle of May, and having
chosen some of the poor, hilly, near cane-
pieces for this purpose, or some of the
bottoms which were poor and gully, on
which water cannot lodge ; I would on the
poorest piece, pitch two of these moveable
pens, line the ground out for them at seven
feet distance for the posts, secure and
tighten them well with strong rails and
stockadoes ; supply them amply with feeding for the cattle ; cover them with all the
dry litter, old grass, cane-tops, weeds, &c.,
which can possibly be gathered and thrown
into them ; pen the cattle in them every

evening, and move them from place to place, where strengthening and renovating is required; taking care when they are moved, to fill up the post holes. Cover the manure with six inches depth of good earth over the surface. When fermented for some time, and beginning to corrupt and rot, let it be chopped, heaped up, and covered with trash. The weather from the latter end of May to September, is generally hot and dry. There should always be a reserved guinea grass piece or two, kept for the purpose of cutting, and feeding the pens. The gullies under shade, yield at this time abundance of assistance to the grass pieces in feeding the pens, and these should be preserved to come in at this period. In dry weather, these pens need not be moved for a month. They will be the better for it, as the greater the quantity of stuff thrown in them, which being pressed down by the weight of the cattle, and dunged and urined over by them, will sooner ferment, and accumulate a good

heap of manure, when chopped and made
up. But from September till the end of
December, the weather is moist and rainy.
The pens at this period will become swampy, from the frequent treading of heavy
stock, and should be moved every two or
three weeks, so that the cattle may be kept
comfortable without lying too damp. Yet
they should remain sufficiently long, to
cause them to tread, mix, and stir up the
stuff in the pen. The grass pieces which
had been rather parched in the hot months
of June, July, and August, assume in those
months their wonted verdure and luxuriance, affording abundance of fine grass
for the cattle by day, (which every good
planter should strictly attend to, and have
driven to the best of feeding) and a plenteous suppering for them during the night.
As a good deal of what manure is made
at the works (which is generally of the best
quality) can be easily carried out to some
of those near cane pieces, it will, in a great
measure, assist in giving the fall plant

rather a greater share of such warm, invigorating stuff, than the spring plant ; as the land at this cold inert time of the year will demand it, to throw life into the plant which lies imbedded in its bosom, foster its exertions, till it embraces with its roots its natural mother earth. I hope I have defined with clearness, the great utility of having plenty of working and breeding stock upon a sugar estate, and the admirable good consequences ultimately to result by its adoption. By such a plan, moveable cattle pens can be formed, where abundance of manure may be made on the land, without carrying it out in carts, and on mules' backs. A supply of this valuable material in plantership, should never be wanting, either from the works, or the pens, to keep up the strength of rich lands, or renovate and restore the impaired stamina of poor decayed soils.

I cannot dismiss the subject of a sugar cane fall plant, without suggesting a few hints, relative to the necessity, of having a

field of canes growing, to serve as a nursery to draw fall plant tops from, which is
a primary object to establish, and have
resort to, in any emergency when the tops
which grew upon pieces lately cut down
are either bad or arrowed. It is essential
to be more particular in picking tops for
the fall, than the spring plant. For if any
indifferent, sour, blemished, or arrowed
tops, are put into the ground in the fall of
the year, when the ground is chilly or
spongy from rain, not one in four will perhaps sprout, but lie dormant and rot,
giving endless trouble in supplying the
piece ; which, when finished, after infinite
labour and expense, will shew an unequal
appearance, of high and low canes. The
arrowed top is a pithy, almost hollow, sap-
less trunk, with little life, few eyes, and
those mostly blind ones. How can such a
plant be depended on, to raise a stock or
root of canes from ? As to sour and blemished tops, planted at this time of the year,
a cultivator might as well put so many

withered bulrushes into the ground to raise cane sprouts. I would always then, have this grand reserve of a nursery, in some secluded place, to be called in as a magazine, (if it may be so called,) to give efficient plant tops for fall plants, in case no other good ones can be had, contiguous on the estate, to answer the wishes of the overseer, to plant out the piece in proper time. I have known some select the worst piece of land on an estate, for the purpose of a nursery. I would act contrary to this, and appoint three or four acres of good ground to be so occupied, in some secluded spot, not to mar the appearance of other cane pieces, by shewing its irregular cuttings. For in a nursery may be observed several growths of canes, making a striking singular appearance, owing to the necessity of drawing plants from it at different times. I would likewise choose cane tops of the most approved species, which may answer the climate and land of the parish, to put in as plants in the nursery. Be particularly careful to keep it clean, and not allow the breed of canes to degenerate, by neglect or trespass. I would always manage it so, that it should not be a prey to arrowing, by keeping up the soil with manure, and cutting it down about the first of March. By so doing, the canes will be of considerable height, and will not be attacked or swelled by arrowing in the October following, when their services will be wanted. I would then take three lengths of plants from each cane, those which are vigorous and lively, and have plenty of full, prolific eyes. These seldom disappoint the well-meant zeal of the expecting over-seer. I knew an overseer in a temperate parish, so fond of thick planting, that he expended sixty acres of tops, in setting out eleven acres of land, and after all this most extravagant work and waste, was obliged to beg plants from his neighbours to supply the piece, or in a great measure to replant it, for want of this useful nursery at this time of the year.

Hilly land on account of its various inequalities, requires the attention of the skilful planter more than level ground, to lay it out in roads, and lining for cane holes. But being steep or of sufficient inclination, it does not want so much trenching to carry off rain and moisture. It has two disadvantages, that it is more liable to be broken into chasms, the banks tumbling into one another, and the land sooner wearing out by heavy rain, than level ground. Yet with all these dispositions to prepossess people against it, nature has thrown it in such a form in the planter's way, that he must take it for better or worse, and turn his mind sedulously to work, to obviate its defects, and improve its deformities and imperfections. After feeding and clothing it with manure, he must then lay it out for cane carriage, and for cane holes, having his large wooden square, level, lines and pegs ready for so doing. He should lay out his winding road so, that it may not rise abruptly, or take up any unnecessary space;

it should be made to serve for a general
travelling road, for negroes and stock
through the piece, to carry off and on the
canes, that a second road may not be required, thus occupying the ground which
should be in cultivation. I shall here beg
leave to quote an instance of this needless
display of art, in having a multiplicity of
roads through hilly cane pieces, with no
other apparent view but to excite the attention of the traveller, his praise might be
resounded to this road-making overseer,
without knowing whether he was right or
wrong in so acting.

The estate I here allude to, lies in the
parish of St. Mary, with a ridge on it,
separating two rivers, which bend their
course north-west, but form a confluence
about three miles below the estate. From
thence their united stream, flows over a
rocky channel to Rio Nuova Bay. This
overseer had likewise a great inclination
for small, trifling cane pieces, insomuch,
that an old cane piece, or a new piece of

ground, was sure to be mutilated by an extraordinary number of intervals and roads, perspicuous without utility, and confounding the cultivation book of the estate, with a list of new fangled and complimentary names. Two cane pieces which lay on a declivity, over some bottoms, nearly opposite the overseer's house, were to be planted. They were thrown out to rest for some time by the former overseer. One contained ten, the other about six acres, and had the remains of former well laid out roads upon them, for mule carriage; fully adequate to the crops, with as much ease and safety, as the nature of the ground would admit. These intervals and roads were left open, and the stock naturally bent their way to them : a sure sign of their being of easy inclination for the beast to carry its burthen. It then became apparent, that making extra roads, was only wastefully cutting up the land, and throwing the best negro labour of the estate away, which should be expended on some more necessary work. The cane piece of ten acres, he intersected with three additional roads, and that of six acres, with two additional roads. At the same time making them so narrow, that when the canes became high, they were covered. In some places of these roads, where great labour was used to dig down high banks, the banks gave way, overwhelming the canes, breaking them up by the roots, stopping that part of the new useless road, and forming galls, ugly and dangerous spots on the cane piece. At last, when the canes were cut down, it was discovered, that little or no way had been made by the stock through the tracks. That the mules instinctively moved through the pieces, by natural easy passages, to the old road, and that the new roads, made so deranged an appearance, that they were stopped up and dug into cane holes. This was an experiment practised by a self-taught overseer, who happened to obtain a situation from the resident agent, on account of some interest he had with the proprietor of a certain estate in the parish of Clarendon, and when that interest was used with the resident agent, favour was shewed to this overseer. His faults however became glaring, his services were no longer necessary, and he was obliged to seek an asylum, in the auspices of another resident agent, who had likewise the same views with the former one.

A road made through a cane piece, for the purpose of easy carriage and conveyance, even for mules, should never be less than eight feet broad, to give room for the mules loaded, to pass and repass one another, without injury or impediment. It should not be covered or hidden by reclining canes. As I hinted before, one well laid out serpentine road, of such a breadth, through a hilly cane piece from bottom to top, will be found adequate for easy carriage. If possible it should not rise more than one foot in nine. A moveable light triangle, with a radius fixed to it, on which

are marked degrees of elevation, and a
plumbet, to move or swing to those degrees, (according as it is wanted to rise or
fall the road) should always be used in
such an undertaking. High stakes should
be driven into the ground, through the
parts of the piece which it is designed
the road should run at certain distances
from bottom to top, as land marks, to direct the attention to, from the point of
departure. And level from that departure,
either at nine or twelve feet on the radius,
towards the first of those directing stakes,
a peg should be put down every three or
four feet ; till it arrives at such stake, and
so on to the next directing stake; levelling
with the radius, and pegging the ground
at every three or four feet, till it reaches
the summit of the piece, or where the road
is to terminate. Then place the great gang
with hoes, pickaxes, and iron crows, to dig
from those pegs, a road eight feet broad.
The mold dug out from the top of the
bank, should be spread on the lower part

of the piece, adjoining the road. Any stones
raised from it, may be placed firmly at the
lower edge of the road, and the road have
from the top bank a gentle slope, or lean to
the lower edge of it, with here and there a
small shallow trench to carry off the water
after heavy rain. I would let as little loose
mold remain on the road as possible, so
that it may soon get hard and durable.
After having made this general road, in a
winding manner through the piece to a
given point, what few trenches may be
wanted, can be laid out and cut, and the
ground cleared of grass and cane roots,
previous to lining it into cane holes. I must
next propound my design in lining a hilly
piece for cane holes. The situation of such
a piece should be well considered, as to its
inequalities, its turnings, and its gullys.
For such a work, there must be a large
wooden square, of four feet, or four feet
and a half on each angle, to lay out the
rows from the interval, or centre of the
piece, and measure the distances of the

rows, or breadth of the cane holes with. Besides which, there must be an extensive pliable line, flagged or marked every four feet, or the width of the cane holes; two or three mule load of strong wooden pegs, to mark the rows with, and a number of high slender stakes, to take an observation or direction from. According to the windings or turnings of the hills, I would divide the cane piece into so many breadths, putting a long stake down at the top, where the turning makes an angle, and another stake at the bottom, where the turning forms another angle. Then having stretched a head line down the steep interval, which separates that from the adjoining cane piece, and close in with the head row of the piece, which head line, when stretched, should be pegged off at every four feet, or the width of the cane holes, to the end of the line, I would either begin to line the cane holes from the top or centre of the piece, but generally from the top, the head row being marked out by the heading line;

or I would take up the wooden square, and
laying one side of it close and even with
the head line, move it by such line to the
top interval, where I wished the cane holes
to commence, and taking the flagged line,
to which a peg should be attached, strike
such peg in the interval, close to the edge
of the square, and extending or stretching
the flagged line by the edge of the square,
and top interval, across the cane piece, so
far as the first tall upright stake, attach
another peg to it. Having tightly stretched
the flagged line by the square and top interval, across the piece, to the first tall
stake, I would then fasten it in the ground ;
and put down a peg at every flag of the
line, where it is fastened in the ground.
Then taking the head line, I would stretch
it from where the flagged line is, at the top
interval, setting the edge of the square
to that flagged line, and stretching the
head line by the edge of the other angle
of the square, down the hill, and parallel to the other head row, marked out

on the steep interval, would put the peg
attached to it fast into the ground, and
then take the measure stick or square
of four feet, or four feet and a half long
and peg off this head line, the same distance
for the cane holes, that the head line was
pegged off on the parallel steep interval.
This being done, I would take up the flagged line, and put it down, and stretch it
tight, between the parallel pegs of each
head line, across the piece ; peg this row
off, take up the flagged line again, stretch
it between the next two parallel pegs, and
then peg this off, and so on, till the breadth
is lined out. Then I would begin at the
top of the piece again, stretching the flagged line by the top interval, to where the
next directing stake is placed, which I
would peg off, and from that stretch another
head line down the hill, by the edge of the
square, pegging this off with the measure
stick parallel to the pegs upon the angle of
the first turning of the hill, and taking up
the flagged line, stretch it tightly between

the parallel pegs, peg this likewise off.
Thus I would proceed, moving the flagged
line between the parallel pegs, and pegging
off the rows by it, till the second breadth is
finished or lined out ; going on, till the
lining of the piece is completed.   Care
should be taken to have no half rows in the
middle of the piece, but if any should
happen, to cause them to come out at the
top or bottom interval.   It is difficult to
instruct some young men in this exercise of
their duty, though they should be proud to
attain it, as it forms one of the prime ac-
complishments of a good planter.   There
are many overseers, who know when a
cane piece is well lined, but are diffident in
attempting the work themselves; therefore, on a great many estates, this part of
the duty is performed by an experienced
slave, (with an assistant to hold the line,)
who has long been accustomed, by the ob-
servance of some former example, to lay
out land in this way.   Bottoms are much
more easy to be lined, as the square, used

with precision, will tell well and true there,
but only to set off the head rows and intervals correctly, which will lead to a general
and fair finishing of the work. I would
never line hilly land at a greater distance
than four feet cane holes, because the
banks are so liable to tumble, the land
being much loosened, and apt to run into
ruts and gullys by wide digging. Hilly
land is more exposed than bottoms, to air
and sun, and does not require such great
distance in the rows, for the air to circulate
and sun to penetrate through them, to ripen
the canes, as bottoms do. The land is
likewise generally poorer on hills than
bottoms, sooner gets worn out by heavy
rains, which wash away some of the best
soil, and often open the ground into dreadful chasms, sometimes carrying it away
altogether. It would be adviseable, then,
where the land is rather steep, to bind it
here and there, by digging check, or
chequered holes, which can be marked out
on the flagged line, pegged off when the

piece is lining, and the negroes, when they are digging the piece, observing where to leave the checks, or small undug spaces. In the cane holes of these steep places, good strong manure should be thrown to help the land, and throw up a strong cane. Those spots are generally poor and gully, and disfigure a cane piece, and do not yield equal to the rest of the piece, unless so attended.

I shall now beg leave to offer some remarks on the difference of turning up land, for cane cultivation with the plough, and that of the hoe. Although in many instances I have thought fit to condemn old customs, which had little to recommend them but stubborn practice, yet I will here incline to give my vote for turning up land with the hoe, in preference to that of the plough, and by adducing a few reasons in support of it, to throw the disparagement on the side of the plough. First, the land being loosened by the close turning up of the plough, gives the sun a

superior influence in absorbing its enlivening prolific powers, for it penetrates with
rapid violence through the soft loosened
stratum, and before the banks are made up,
or a gang can be set in to raise them to
regular distances by the hoe, the parched
earth loses most of its stamina, by moulder-
ing into a crumbled, a dry, wasted, spent
substance. Secondly, the people in that
country are most ignorant of the art of
ploughing; the cattle and negroes are
hard to be trained to it; ploughmen sent
out from home are at a nonplus for what
they think are requisites; much time is
spent in getting every thing to their liking,
and when they do set to work, the cattle
are restive, and the negroes refractory; the
arduous task begins too much to be borne
in that hot climate, by the white ploughman; he becomes tired, listless, and perhaps sick, before the piece is finished; and
if it is finished, it has taken an immoderate
time to do so, (even longer than if dug by
the hoe,) and all this time the loosened soft

earth is exposed to the parching rays of the sun, to dissipate its strength. Thirdly, after the piece has been turned by the plough, a gang must be set in with hoes, to haul up and form the banks regularly, and clear the cane holes for planting, which is extra work. For if the piece is turned up by the hoe, this work is done at once, the cane holes are properly dug, opened, and the banks hauled evenly up, and every thing left in *statu quo*, to commence planting the piece. The labour, and perhaps the lives, of the best cattle on the estate will be saved, by not adopting the plough in this work ; time will be saved in getting the piece planted, to catch the favourable seasons ; the strength of the soil will be preserved by not being too much loosened, and the salary and maintenance of a white ploughman will also be saved. The work being gone through with, and perfected by the hoe, the implement mostly used, and best understood in that country, will be timely and satisfactorily completed, to the advantage of the estate, and the good understanding and comfort of the white people and slaves upon it.

By digging with the hoe a piece of land into cane holes, there is always some part of the ground left undug, that is, the base of the banks. For although the cane holes may be lined four or five feet wide, yet the bottom of the cane hole is not near so broad as the distance between the top of each bank. The banks are hauled up to the forming pegs with the hoe, in a sloping manner, making the bottom of the hole not more than sixteen or eighteen inches broad; whereas the distance between the top of each bank may be four or five feet; so that when the canes are planted, and the banks are levelled to them, the distance between each row of canes will be three or four feet; giving liberty to the negroes when set in to clean, trash, or cut them, to have room to work without injury to the canes, and the sun and air to penetrate through to ripen them. Another matter to specu-

late upon, which I believe few have thought
of or attempted, is (by a simple process) to
gain fresh ground for digging, the next
time the piece is to be turned up for cane
holes. Having before stated, that the base
of the banks, in digging cane holes with the
hoe, is left untouched or undug, it surely
must strike the attentive observer, that
such base is virtually fresh ground, or
nearly equivalent to it, so that when this
cane piece is to be dug again, in order for
replanting, by either narrowing or widen-
ing the interval so much as the breadth of
this base or undug ground, fresh strong
land to dig and plant in is gained; and the
former space between the rows of canes
becomes, by such a plan, the ground in
which the subsequent canes will be planted.
I will not, however, in adopting this simple
plan, any way derogate from the maxim
and system which I so strenuously recommended for keeping up the strength of cane
land, by rich, and the never-failing art of
manuring. This little alternative will give

T

relief to the space of ground formerly occupied by the canes for their support; it
will give it time, in a great measure, to regain its strength, which has been much
exhausted in raising a crop of plants, and
perhaps three crops of ratoon canes from
the same spot, or four years and a half of
unceasing cane vegetation. But while it
thus gives relief to the old occupied spot;
while it thus supplies fresh ground, to encourage the growth of fine strong plants,
the never-to-be-forgotten plan of manuring
should be carried on with all its force and
effect. The new ground should have its
full allowance of enduring manure, while
the old ground may retire within itself, to
give nature an opportunity to recruit it, by
a rest for some crops. Any soil the old
ground loses, in superficially drawing, as it
were, some of its mold, to give the adjoining ratoons occasionally, (which may only
happen twice a year,) will be amply made
up by the rotting cane trash which is pulled
off the adjacent canes when they are trashed,

and the piece is cut down to make sugar, or when trash is turned on it; and which is always regularly deposited in the spaces underneath which the ground lies, which is thus consigned to rest; instilling into it gradual nourishment, and adding to its new powers.

Hilly or distant cane pieces having been prepared in the manner I have prescribed and laid down, by manuring, trenching, clearing of grass, weeds, bushes of every description, stocking up of cane roots, &c. and lining with true method the ground for cane holes, I would propose to set in the great gang (first invoking Providence for favourable weather to begin, go through with, and end such work,) well appointed, with strong, good-sized, sharp hoes, and with heart and hand begin digging the piece into cane holes that is first intended to be planted. If the breadth is sufficiently wide to admit the entire gang to be spread, so much the better, as they not only perform their work to more advantage by being in a regular unbroken breadth, but have a more sightly and better marshalled appearance. Their relative strength, one with the other, is better observed, equalised, and apportioned. Besides the piece is sooner dug by this regulation. However, if it happens that the entire gang cannot be ranged in a regular breadth, let the gang be split or divided into two bodies; one of which should be placed at about fifteen or twenty rows below the other, that the driver and bookkeeper may have them nearly in view, to observe and make them do good work. When the top division finishes digging to where the bottom division began, they should be moved down fifteen or twenty rows below that, and so on as often as possible, till this breadth or angle of the piece is dug out; then rise to the next breadth, digging each down in like manner, till the piece is completed. Care should be taken, as they dig the ground, to make them haul the mold well up to the forming pegs. It should lie firm in the banks, to clear the bottom of the holes of all loose mold; they should dig deep, from eight to twelve inches beneath the surface, even with the pegs, and the banks must be hauled up well, and straightened with the pegs so as to exhibit an even uninterrupted row. A gang of forty able negroes will dig an acre per day of cane-holes, when the land is not too hard or tough, and the weather moderate. At this laborious juncture, pride, which in some measure is inherent in all mankind, animates the soul and body of the negro with emulation. He is associated with tried veterans of equal powers and talents in the exertion of his strength; he bestows his pains with exultation, to benefit his owner, and please his overseer, and driver, and by so doing to gratify himself. No wonder then, when his best exertions are called forth, his strength put to the test, his mind with ardour roused, and with stimulating vanity urged to the work, so that

none shall surpass him, and no fault be
found with him, I say, no wonder, that some
little recompence should be given to him,
some little reward for his pains, to refresh
his exhaustion, prevent him from catching
cold, excite his powers, and sustain him under the fatigue. At this period it is found absolutely necessary to give each negro a certain quantity of rum and sugar, not so
much as will make them drunk, but sufficient to keep their spirits from flagging,
and prevent if possible the intrusion of cold.
The sugar being diluted in water, with a
mixture of limejuice, is found to be both
refreshing and nutritious for them, and they
should never be denied these salutary,
grateful restoratives, when engaged in the
laborious, arduous undertaking of digging
cane holes.   An addition likewise of salt
provisions daily, should be allowed them
when so employed.   There will be found
no loss in it, for though ignorant and barbarous, they are extremely sensible of
favours and rewards, and made amenable

17

by such simple, such trifling, cheap gra-
tifications.

It is a common practice in planting the
sugar-cane, to let the land remain unplant-
ed for four or six weeks, to induce the
banks, by being exposed to the weather, to
pulverize. This speculation comes from
the old school, so much noted for dictating
rules, without much reason for them;
thoughts coming suddenly into the head,
or from some philosophical hints, which
they transiently have heard from the mouth
of some occasional visitor. These they
eagerly have caught at, treasured up, and
followed (as they thought) with infallible
devotion, and certain success. In my remarks relative to plough and hoe turning-
up of land, I made bold to descant on the
bad effects of having loosened land exposed
to the influence of the parching rays of the
sun; for though that luminary is benefi-
cently kind, it is armed with fatal burning
shafts, which the Great Father of the universe bids us be prepared and guarded

against, that while we court its favours, we
should studiously avoid its killing powers.
Those old dogmatists in this case, very seldom have taken into their consideration,
what a long continuance of dry, and sometimes wet weather, happens in Jamaica,
even in the temperate parishes; and that
by exposing land in this way, for their for-
tuitous schemes, they, ten to one, fail entirely of the wished-for pulverizing result.
That by a month of scorching hot sun, the
banks or clods are rendered harder, any
stamina that might have been in the soil is
absorbed and expelled, and by exposing it
to incessant rain, a great part of the banks
are levelled, broken away, the bottom of
the cane holes swamped and chilled with
water, the grass and weeds grow with
rapidity and luxuriance, scarcely to be re-
pressed, both in the cane holes and on the
banks; that require a strong gang of negroes to clean them, before the cane-holes
can be planted with tops. And what is
worse, the land gets no way enriched, no

way improved by such exposure; on the
contrary, the sun cannot enrich it by burning, with its hot penetrating rays, nor the
rain by swamping, and carrying away its
best particles of growing virtue. It would
be much better then, not to depend upon
the casual event of this parching, swamping, pulverizing scheme, but to plant the
land with cane tops as early as possible,
after digging it, and have the virtues of the
soil imparted to the plant, with as few accidents of this nature as possible, or when
it is possessed of its native vigour, and when
it is fresh; giving Providence a paramount
right to have a sway over it, by taking advantage of the earliest opportunity of time,
place, and circumstances, to get the plants
in the ground. The seasons must not be
suffered to pass by, that are propitious for
planting; take then and plant the ground
out while it is fresh, if it is possible to procure good tops for so doing. The seasons
that the pulverizer waits for, will be more
favourable for bringing up, or making the

plant shoot or sprout, than for forwarding
his notions. If the ground is to be melted
by rains, those rains will induce the plant
to grow, enough of light soil will always be
found about the cane holes, adequate to
cover the top with when planted, and when
succeeding care requires, fresh mold may
be added to them. The banks then will
be soft, free from weeds, and a proper
supply of mould can be given them.   I hope
I have explained this matter, to the satisfaction and conviction of my readers.

Having the plant tops carried to the contiguous intervals, the great gang is to be
set in to plant the piece, attended by the
active weeding gang, with dung baskets, to
dig, and drop manure into the cane holes
before them. If the manure which has
been made by the moveable cattle-pens is
too much to spread on the ground, what is
left should be carried out to the interval,
heaped, covered up, and left there till
wanted to give the plants their second
dressing. The great gang having stripped,

trimmed, and carried as many good fresh
cane tops to the holes as will plant out a
breadth, should set in to plant, placing a
negro in each row. They are to spread the
dung thickly in the bottom of the hole, first
cleaning it well out, and weeding the banks,
which weeds and grass must be shook well
from any earth adhering to them, and their
roots turned to the sun, to wither and kill
them, placing the weeds on the top of the
banks. I would make them plant the tops
across the holes, in the hilly land, that they
may not be so liable to be displaced by rain
or accidents; and although I am no friend
to thick planting, I would do it rather
closer than bottoms, the sun and air having
greater power on it. I would then make
each negro take three good plants, and
sprinkling the dung over with earth, place
firmly in the bottom of the cane hole, on
the soiled manure, at eight inches asunder,
the hard end of one top opposite to the
furry end of the other, and then cover them
lightly with mold. Each negro being

provided with a measure stick twelve inches
long, should measure with it a distance or
space, from the most forward of the three
tops last planted, and having the manure
spread and soiled as before, plant three
more tops, at eight inches asunder, which
likewise he should cover lightly with mold.
Taking the measure stick again, he should
mark out another distance with it, spread
the dung and soil as before, take three
more good tops, plant them firmly, at eight
inches asunder, cover them lightly with
mold, and so proceed in planting out the
piece till it is finished ; taking special care
to weed the cane holes and banks well, as
he goes on planting, and place the weeds
on the top of the banks, with their roots
turned up. The trenches likewise should
be well cleaned and cleared as the gang
plants the piece; the surrounding fences
trimmed, cleaned, and made up. The intervals cleaned of weeds, all grass hoed off,
six feet distance from the cane plants, and
it should be billed down to thirty feet dis-

tance from it ; so that rats will not have an
asylum close to the plant canes. Indeed,
the last direction is a most necessary one :
it not only may be the means of preventing
a great deal of destruction to the cane
pieces adjoining, by the depredations of
these lurking gnawing pests; but where
these places are cleaned, guinea grass can
be planted and cultivated, which at all times
will be of essential use in feeding the cattle
pens and mule stable. The grass cutters,
as they cut the grass daily, can keep it clear
from bushes, and the rats will seek safety
in a great measure, by retiring further back.
As I explained before the utility of planting corn (for the sake of plenty) over and
through cane pieces, I will here again venture to support that measure, by planting it
on the side of the banks, at five or six feet
distance. No injury can happen to the
canes by so doing, it being a hasty crop,
and gathered in four months after planting,
in favourable weather. It has been known
even in very hot weather, to have been of

service to the plant canes, by shading them
from the scorching sun ; I would then give
it always a sanction, having a great many
virtues, and but few vices. It will be unnecessary here to repeat the mode of weeding and moulding the plants as they come
up, till they require the banks to be levelled
to them. This I trust I have sufficiently
explained in a former place, where I dwelt
on the method of planting canes on bottoms,
weeding, manuring, and molding them,
till fit to have the banks levelled to them ;
to which I beg leave to refer my readers.

Much depends on the kind of weather
which happens, or the soil it grows upon, to
instruct the planter when to trash the canes.
In dry weather parishes, such as St. Cathe-
rine's, St. Andrew's, Kingston, part of St.
David's, St. Dorothy's, part of Clarendon,
part of St. Elizabeth's, the west end of St.
Ann's, and Trelawny, one trashing is sufficient after the banks are levelled, for there
the sun and the soil together, in a great
measure, combine to dry up the juices, and

ripen the canes at an early age. Plant
canes in those parishes will be fit to cut at
twelve months old to make sugar ; and they
are obliged to adopt a hardy callous species
of cane to plant there, called the riband,
and transparent cane, in order to withstand
the force of the sun, and the light weak
soils of those districts. They must manure
high, the cane will grow exuberant, but not
yield the quantity of liquor they do in temperate parishes. They give a great quantity
of trash from the mill, even more than adequate to boil the sugar, which is frequently
thrown into the cattle pen to make manure,
but often produces neither so much, nor
such good sugar as the yellow, long jointed,
thick Bourbon cane. This last, and the transparent cane, is what I would promote the
planting of in most parishes, where I could
find strong soils to support them, for as to
the action of the sun upon the cane, that in
a great measure can be obviated, by not
trashing much. The Bourbon cane is a
cleaner, longer jointed, and more productive

plant than other species of sugar canes, and by giving it a little more age, together with due attention in keeping it clean, it will yield more, and better sugar than other canes. Moreover, it always stocks better, and lasts longer in the ratoon, if planted in deep strong ground, with adequate manure. In temperate parishes, such as St. Thomas's in the east, part of St. David's, Portland, St. George's, St. Mary's, the east end of St. Ann's, part of St. Elizabeth's, St. Thomas in the vale, and St. John's, I would always plant the Bourbon cane, and trash them according as the weather would allow. In the more moist districts, that is St. Mary's, St George's, and Portland, I would contrive to give them three trashings, after the banks are levelled to the plants, before they are declared fit for the bill, or to cut to make sugar ; and two trashings to the ratoons, after they are six months old, before they are cut. In the other temperate parishes, two trashings to the plants, after the banks are levelled, be-

fore cut to make sugar, and one trashing to
the ratoons, after they are six months old,
before putting them down for sugar. But
those operations of dressing the canes, to
bring on maturity and good yielding, in
such temperate places, must be considered
as casual; for the influence of weather
should direct this event. Hilly lands do
not require so much trashing as bottoms,
but the canes are more liable to recline,
and lodge on those rising grounds; and
whenever they do receive such dressings,
care should be taken to place the trash in
bolsters under the recumbent canes, to prevent them lying on the bare earth, their
joints taking root, throwing out early
suckers, breaking off from the roots, and
vitiating their juices. Every time the cane
pieces are cleaned by trashing or otherwise,
the gang should uniformly clean the intervals and trenches of weeds, or any thing
else which impedes or injures the canes in
their advance to maturity. I have now
brought this part of my system to a state of

U

forwardness, preparatory to commencing crop on a sugar estate in Jamaica.

I must beg leave to draw the attention of my readers to the establishment of fences on a sugar estate, principally to those for the security of the cane cultivation. Few intersecting fences are wanted between cane pieces, except where a line fence between the canes of two neighbouring estates may be deemed necessary, to mark the boundaries or prevent trespass. The girding fences are those which claim attention, to prevent intrusion from roads, outlets, and pastures, into the bosom of the cane pieces. These defences should be so established and kept up, as to form an impenetrable barrier against cattle, mules, pigs, or sheep ; and at the same time, not to allure (as an asylum) the crafty, hungry, and destructive rat, to take up his quarters there. For this purpose, I should always prefer quick-fences. As to the erection of stone fences round a cane piece, I esteem them as ever truly pernicious and tottering.

They are strong habitations for rats and other vermin, which they cunningly retire to, being attracted by the vicinity of plenty thus to satisfy their voracity. In those walls they form an innumerable society, and propagate with amazing fertility. Some artless, superficial overseers, have a great liking to stone walls about cane pieces, even in the neighbourhood of river courses; the haunts and lurking-places of the vagrant rats, as if charity actuated them to provide meat and drink at the same time for their most inveterate enemies. They would never be out of employment for the mason slaves, in repairing those loose walls, after every heavy fall of rain, gust of wind, or pressure against them; yet I have known some overseers persist in this destructive plan, though the evil of it was glaringly striking to them; because they thought that kind of fence looked neat, resembled inclosures in the mother country, and was durable and forbidding to cattle; not for a moment considering, that they being loose

stone walls, without mortar to keep them
together, they were ever tumbling in some
quarter. I suppose also that such overseers,
know very little how to plant, or establish
a quick-fence in Jamaica. They are
ashamed to shew their ignorance, impatient
to produce something worthy of notice, but
loth to wait for the gradual, yet durable
and beautiful production of a well-esta-
blished quick-fence. Nothing is more
ornamental, and durable on an estate, if
well attended to. The logwood and lime
quicks are mostly planted in Jamaica. For
this purpose, in the proper season, when
the seed of those trees is ripe, (from August to the end of October), there should a
quantity of it be gathered, and dried in the
sun. A small piece of ground may be dug
up and prepared, in order to sow the seed,
from which to draw plants; or there may
be a drill lined out, where it is intended to
put down the fence, which should be dug
about four inches deep, and eight inches
wide. In this the seeds are to be sown,

but not too thick. Previous to digging the drill, sowing the seed, or setting the plants, there should be a durable post and rail fence made, on the outside of the place where the quick-fence is to be planted, not only to protect it in its infancy, and to establish permanent high growth, but to ensure safety to the adjoining cane pieces, from the trespass of cattle. When the fence is put up, let the drill be lined out, and dug in the inside, according to the turning of the road, or the formation of the cane piece; keeping about four feet distance from the edge of the canes. If there are logwood or lime plants to set or put down, let their tops be first nipt off for an inch or two, then be set six inches asunder in the drill, in a zig-zag manner, making a hole with a setting stick in the drill for each plant; into which introduce the root end of the plant, as deep as the fibres of the root appear on its stem. Take care that the stem is some inches above ground, and press the earth (but not too hard) about the root. After the plants are set in this manner, draw round them, into the drill, some of the earth that was dug out of it. The plants will soon strike root, and when that is the case, give them more mold, and let them be kept clean from grass and weeds, and neatly and evenly trimmed. Care should be taken to let no stock, of cattle, mules, sheep, or pigs, touch them, till they are established. As they grow up, preserve them close, with their branches horizontally, and transversely interweaving each other, so as to form a compact, firm barrier, when they are about five feet high. This should be their standard, and they should be kept lopped or trimmed to it. A logwood fence in my opinion, grows quicker, inclines to closer contexture, is more hardy, durable, and firmly resistible, than a lime fence. It is equally as beautiful to the view. It resembles, in a great measure, the white thorn fence of the mother country; and when established, requires cleaning and trimming

only twice a year, in February and August.
It harbours no rats or vermin, withstands
tempests and rain, gives adequate security
to the cane pieces, and highly embellishes
the estate. If it is intended to sow logwood, or lime-seeds in a drill, for a fence
round a cane piece, it should be dug five
inches deep, and eight inches wide. The
mold must be hauled well out of it to the
edge of the drill, broke fine, and made
a bank of. The seed should be dropped
along the bottom of the drill in a zig-zag
manner, but not too thick, and covered up
lightly with mold; it will soon shoot up
and bud sprouts, which when two or three
inches high must have more mold; and if
found to come up too thick, the superfluous
ones should be pulled out. After this they
should receive the same kind of treatment
as the fence put down from drawn plants.

The bamboo cane is a great acquisition
upon every estate, not only to make and
support fences for rails and stockadoes, but
as fuel for the distilling-house. When a

property happens to be either divested of
the article of copper-wood, or the woods
lie at a great distance from the works, they
are excellent fuel, and if planted at a small
distance from the works, they save a great
deal of negro and mule labour, in cutting and
carrying home wood from a distance ; and
when established they are inexhaustible,
for a new, strong, and luxuriant growth will
rapidly spring from the same stock every
year. Two or three acres of them, planted
even in indifferent ground, and taken care
of for twelve months, till they begin to stock
well, with a fence round to protect them
till that age from the intrusion of cattle,
will, after they are two years old, give an
ample supply of copper-wood annually for
the distilling-house, and abundance of long
stout rails for fences and stockadoes. So
little do some overseers know the value and
utility of this very beautiful, shady, and
profitable production of nature, that few
plant them ; and others, when they have
been planted, and half-reared by their pre-

decessors, suffer the cattle to go in among
them, eat their tender leaves, and tread the
young stock out of the ground. It is sup-
posed that there is no vegetable production
in Jamaica which grows so quick as the
sucker from the bamboo stock. They often
shoot more than twelve inches in twenty-
four hours, and in a short time acquire a
degree of hardness scarcely penetrable. It
is hollow, and chambered within ; some of
the chambers from joint to joint will be
two feet long, and may contain a couple of
quarts of liquid. It soon gets dry when
cut, and assumes a beautiful yellow outside,
but will burn equally as well when green,
for nature has given it an oily texture.

It should not be planted close to a cane
piece, because the roots when established
spread to an amazing extent in length,
breadth and depth of ground. It over-
shadows the canes, which is injurious to
them, and may harbour an infinite number
of rats, which would be still more hurtful.
For the purpose of planting them, I would

choose a patch of waste land, not far from
the works, but apart from the cane pieces,
of two or three acres.   This I would fence
round, to prevent any trespass in it of cattle,
mules, pigs, or sheep.   If the ground was
foul, it should be billed, hoed off, and
cleared; then let the ground be lined out
twelve feet square, and pegged off to that
distance.   They must be planted wide, to
give space for their roots to spread, and for
the numerous stems which spring from the
stocks, which grow in clusters of thirty or
forty from one root, of a considerable thick-
ness and great height, and require room for
their lofty magnificent stems to wave.
They are armed with long slender branches,
and clothed with long, narrow sharp leaves,
which throw shade round their vicinity.
A gang of negroes should be set in with
hoes to dig good deep holes at each peg,
two feet long, one foot broad, and nine
inches deep.   They should haul the mold
out of the hole to the edge and break it, if
the clods are too large: and having the

16

bamboo plants ready cut and trimmed, of about two feet long each, (and taken from the ripe green joints, towards the top of the stem), the negroes should plant one in each hole, letting them lie a little reclining in it, with a small part of one end above ground; then let them take half the bank and cover the plant with it; if seasonable weather, young shoots will appear, sprouting from the joints of the plant, in a month after they have been put in the ground; and when six or eight inches above the ground, they should have the remainder of the bank of mold given them, and be cleaned from any foul grass or weeds; they require as much cleaning and molding as a piece of sugar cane, till they are twelve months old; then they will take care of themselves, except from trespass of cattle. The surrounding fence should be kept up, to preserve them till they have well taken to the ground, and become hardy, tall, and well established; then they will begin to repay all the attention given to them, by affording a perpetual supply of copper-wood annually, and materials for fences. Every overseer who knows their value, I am sure, will be an advocate for establishing a patch of them on every sugar estate.

Things being left in this forward clean state throughout the cane cultivation, I shall next beg leave to call the attention of my reader, and the planter, to another care which is particularly incumbent on him; and that is, the cleaning, planting, and fencing the guinea grass and common pastures on every sugar estate, previous to beginning crop ; the neglect of such is sure to bring on some heavy loss or misfortune at no remote period. Whatever may be the effective strength of the estate in slave population, this most essential duty should ever compose a part of their annual work. Even twice a year would not be too much to clean those grounds, which are appropriated to raise provender for the maintenance of the cattle and mules, as little can be done on an estate without their assistance. It

would be better to contract the sphere of
cane cultivation, than suffer the grass pieces
to run to ruin, or become an open common,
as a heavy loss and painful disappointment
must be the consequences of neglecting
them. They are easily kept clean, the
growth of nourishing grass maintained, if
regularly once or twice a year they are well
billed, and the sour grass and noxious weeds
stocked up ; but if they are suffered to run
to ruin, the grass wears out, the sun is ex-
cluded from giving its vital succour to the
roots ; the strength of the ground is exhausted, in the support of superfluous bushes
and weeds ; the natural and artificial nu-
tricious grass for feeding the cattle becomes
extinct ; want ensues ; the cattle and mules
exhibit a lean, ragged, starved appearance ;
weakness seizes them, and they are unable
to go through the heavy work that is allotted
to them. Such are the effects of letting
pastures on a sugar estate go to decay.
This is done for the sake of the shallow,
ostentatious motive, of keeping up a large

field of canes, a great part of the product of
which must, in the sequel, be laid out in
the purchase of cattle and mules, which
have been killed by over-work and starvation; or what is nearly as bad, being
forced to employ a jobber, at a heavy
expence, to bill the neglected pastures,
which can scarcely be reclaimed, the disease becoming incurable. Neither (will I
make bold to assert) is any work of this
kind ever done so well by a mercenary,
superficial jobbing gang, as it will by the
estate people. Another matter I must beg
leave to advert to, is the propensity in most
overseers of the old school to try every
means to swell the crop of sugar, at the
risk and expence of every thing else, of
equal magnitude, if taken upon the great
principle of preserving the capital of the
proprietor. Nothing argues well with the
persons but the magnitude of the sugar
crop, which by its sickly bulk brings on a
fatal malady to other component parts of
the estate; and the only thing gained by

it is, that the overseer's vanity is fed, and
his fame spread about, for making a large
crop of sugar. The resident agent seldom
visits the estate, and when he does, saunters
a little in the great house, takes a transient
view of some of the near cane pieces, and
asks a few cursory questions. The distant
cultivation is not often noticed ; and he is
kept ignorant of the wretched state of the
once verdant and flourishing pastures. He
mounts his horse, or gets into his chaise,
and drives off somewhere else, quite satisfied that the overseer and himself have done
their duty, and does not pay another visit
to the property for six or twelve months.
At last the mortality among the cattle and
mules is such (the woeful consequence of
hard work, starvation, and disease), that
the overseer must divulge and make known
to him the sad story. He prays for a fresh
supply of cattle and mules. This astonishes
the resident island agent : he is roused
from his lethargy, hastily comes to the
estate, and there beholds the sad remnant,

the miserable spectacle of the few remaining cattle and mules, reduced to such a
state by intense work, in taking off a heavy
crop of canes, in all weathers, and starved,
by the improvident measure of neglecting
the pastures, in order to increase the overseer's reputation in making a great (but
overcharged) sugar crop, and to swell the resident agent's charge of per centage on the
account sales. He then, to exonerate himself
from any charge of dereliction of duty, ac-
cuses the overseer of imposing false accounts
on him, assumes an air of dissatisfaction, blus-
ters a little, goes home, and seeks for another
overseer to succeed the former. In a short
time he proceeds to relieve him of his charge;
and then the whole history of former mis-
management is developed and explained.

A gang of forty negroes will clean better
than one hundred acres of pasture in a
fortnight, if not too foul. The masons
on an estate should, at the same time, go
round the stone-wall fences, to repair them
where they are broken. This being finished,

(except very dry weather ensues for a long time,) there will be plenty of provender for the stock during the crop, with the addition of guinea grass intervals, gullys, and margins, (which are contiguous to the cane-pieces,) to cut, in order to provide feeding for the mule stable and cattle pens. While the great gang is employed in billing or cleaning pastures, the second gang should be engaged in the cane field, where most necessary. The carpenters should be employed in making the necessary repairs to the buildings, preparatory to crop. The coppers should be employed in splitting staves and heading, in the woods, which may be carried by mules to the works, and heaped and piled up there. After the stone wall fences are made up, the masons should be at the works, making preparations in their department for what may be wanted about the coppers, stills, the different buildings, and white-washing the whole of the houses, inside and outside. The copper-smiths and plumbers should be soldering any cracks or split seams in the leads of the coppers, mill-beds, liquor-gutters, coppers, and stills. In fact, every denomination of tradesmen should be fully employed after the old crop, in providing for the new one.

It may be necessary to enlarge the pastures sometimes, or to put in, or plant another guinea-grass piece, when there may not be sufficient feeding or grazing for the stock on an estate; to fatten or recruit those, that are either condemned as unfit for work, or are reduced from hard work, disease, or sores. I would always strive to keep the grass cultivation as connected as possible, and subdivided by stone walls or double ditches, one from the other, and penguined over. In the vicinity of pasture lands, on an estate in Jamaica, there is generally some waste ground that will answer to plant guinea-grass, as an additional nursing, fattening pasture. It does not require the best land to plant this artificial grass, or a great depth of soil. Ground

that may be somewhat stony, will suit it,
and if it abound with stones, may be cleared
of them, to build a wall round the pastures
where wanted. If a piece of land, of eight
or ten acres of this description can be obtained, close to, or adjoining the other pastures, so much the better ; it will be still
more favourable, if a spring or pond of water
is in it, as the fences of the old pastures will
divide it from them, and partly surround it.
To plant and establish this piece of land in
guinea-grass, I would set in the great gang
with good sharp bills and hoes, and if the
land has any underwood upon it, it should
be cut down, junked, and sent home by
mules to the works, for copper-wood ; then
billed down and hoed off from bushes and
weeds ; no fruit trees, such as lemons, limes,
oranges, citrons, or even large guavas,
should be cut down, except they are too
thick on the ground. When the bushes
and weeds are dry, they should be burned,
and the land being thus cleared and prepared, the masons should be put in, to make

a five-foot wall round it, (of the stones in
the vicinity,) on the defenceless side, to
join the fences of the old pastures, as far as
the stones will go; and if there is not enough,
the fence may be continued, by a double
ditch, topped with penguins, to reach and
join the fences on the old pastures. The
carpenters should make proper gates for the
ingress and egress of the cattle. When the
land, in two or three days, has been cooled
from the burning of the weeds, and Providence sends favourable, showery weather,
I would again set in the great gang with
hoes to dig the grass holes. The ground
should be lined at two and a half feet
square for them. The holes should be six
or eight inches deep, twelve inches long,
and eight inches broad, and the mold
hauled up to the edge of the hole as a
bank. Then having plenty of guinea-
grass roots prepared, trimmed, and divided
into proper plants, of a size which may
easily be grasped in the hand, let the negroes drop two plants at each hole, through

16

the piece, and begin planting them, by placing the two plants for each hole opposite one another, in a reclining position, in the bottom of it, with their roots down in the hole, and nearly touching each other. The stalky part of each plant should be above ground a little, and extend in a reclining way, opposite each other, to the extreme length of the hole. Then the mold should be drawn from the bank upon them. In this manner the ground may be planted out in guinea-grass. The best time to plant it, is the month of May, for it will seed (if taken care of and kept clean) in the following October. Guinea-grass requires to be molded, when the stalks and roots throw out new stalks and grass shoots. The holes should be well filled up with soil, and any sour grass and weeds, which may spring up through the piece, must be well hoed off, whenever the piece begins to be foul. It will require at least two cleanings before October, when it will seed, and when ripe, the seed will scatter through the piece, and sow and establish itself. After the first seeding, some light cattle may be turned into it, to nip the top of the grass, but no more, and their feet will tread the seed into the ground. Then let the gates be fastened, and the piece shut up from grazing or cutting, till the grass is quite high, and the seed has taken to the ground, and grown high.

When the great gang has finished cleaning the pastures, and planting guinea-grass, they should be set in to cut, and split as much copper-wood as may answer to boil one-third of the distilling house crop. The able negro men ought to have sharp axes, and bills, and the women sharp bills for the purpose. The wood should be procured in the most convenient place, for the carrying it home, and lopped to such lengths, as will fit the mules' backs without galling or incommoding them. This copper-wood should be brought home to the works in as favourable weather as possible, the carriage being distant and burthen heavy, so that

the mules may have the advantage of dry
roads, and the burthen made comparatively
easy to them ; and should be piled up compactly near the distilling house.

It may be necessary to build a lime kiln,
preparatory to commencing crop on a sugar
estate ; for this should never be done if
possible during crop.   If no fixed kiln is at
the works, it will be absolutely necessary
to take the great gang, together with the
carpenters, masons, and all the able hands
which can be mustered, and proceed to the
woods, for the purpose of making one, the
men with axes and bills, and the women
with bills and baskets.   It is a most laborious, expensive, and dangerous work.   Some
of the best timbers and wood on an estate
must be felled and lopped to certain lengths
for the purpose ; each tier of wood must
have its different measured lengths, according as they proceed in building the
kiln.   The bottom tier, may be composed
of the heaviest timber, and greatest lengths,
with the thick end of the timber outwards ;

in order to give an inclination to the kiln,
according as it burns, for the lime to fall in-
wards. The masons and coppersmiths
should be provided with sledges, iron crows,
hammers, and picks, to raise and break the
stones small. When wood and stones
enough are cut and broke, to form the bottom tier, a circle of twenty feet in diameter, should be scribed and marked out,
on a spot, where the materials are contiguous, and a high stiff stake drove in the centre of this circle, to form the draft funnel
by. The negro men should then lay the
thick end of the heaviest wood to the edge
of the circle, and the thin end towards the
stake in the centre of it; leaving a small
space for the funnel to be raised from:
when the tier of heavy wood is laid all
round the circle, and close together, and
the interstices filled up with small wood, so
as to make it level and even, then the women
and inferior people must take their baskets,
which they are to fill with the broken
stones, and throw them on the top of the

wood, which the people who laid the tier
of wood should spread on it, at least a foot
thick. An experienced carpenter negro
should be placed at the centre stake, to
build the funnel, as the kiln advances in
height. When this tier of wood and stones
are laid compactly, the axe-men are to cut
lengths of timber, about twelve or fourteen
inches shorter than the under tier of wood,
sufficient for another tier; and stones are to
be broken likewise for it. When this is
done, let them lay the second tier of wood,
on the top of the tier of stones, with the
small end of the timber close to the funnel.
This must be made to lie close, and fill in
the crannies and spaces with small wood,
so as to make the kiln gradually small to
the top, and incline to the centre. Cover
this likewise with about twelve inches of
broken stones, taking care to proceed in
forming and making the funnel, as the kiln
gets high. In this manner keep on building the kiln, for six or seven tiers; gradually making each tier of less diameter, and

putting a sufficiency of good heavy timber in it, intermixed with small wood, in order to ensure the burning of the stones thoroughly. When the lime-kiln is built as high as is requisite, or the before-mentioned number of tiers, let a dry ants' nest be obtained, which being set fire to, must be thrown down the funnel, and some dry sticks along with it, and placed over it, so as to catch fire. A negro should be placed to watch it, and feed the fire at the bottom of the funnel, with light dry chips, till the fire catches the adjoining timber with proper effect. There is then no fear of the lime-kiln not burning well, and in a few hours the whole of the timber will be consumed, the stones burned through, and lying in one promiscuous heap of strong lime. A kiln of twenty feet in diameter, will make sixteen sugar hogsheads of lime, or perhaps more. It will take some days to cool before it is fit to carry home ; and it may be necessary when it gets cool, to make a shed over it to protect it from the weather.

The building of a lime kiln in the woods of Jamaica, should never be done if possible : it is laborious, dangerous and expensive, in drawing off the principal people of the estate from other work.   By having a fixed kiln at the works, a few hogsheads of lime can at any time be conveniently and speedily burned ; there will then be fresh lime for tempering cane liquor whenever required, and a great waste of it prevented ; which cannot sometimes be avoided in the alternative of going to the woods to build a kiln.   One half of it often lies there to be washed away by rain, or stolen by the negroes.   It likewise loses its active alkaline qualities, by exposure to the weather and air.   The carriage home will be saved, which is a troublesome job, for the negroes and mules.

One thing more I shall beg leave to intrude on the notice of my readers and the planter, which is, the care and cleaning of the provision grounds, previous to beginning crop, that were planted or established for the support of the resident white people on the estate. There are many planters who think this duty even worse than a minor object in the scale of true plantership. This is a despicable thought, for if there are no such provisions, if there is a dearth of them, where is their resource for a supply? Is it by begging of their neighbours, or the slaves of the estate, that they are to obtain it? or is it by bartering away the produce of the estate, that they are to be furnished with such necessaries? In any of those cases, there is a lamentable precarious dependance. If we do not sow, we cannot reap; if we do not plant, we cannot gather; and if we abandon those necessaries of life when planted, to perish for want of care, we commit a double crime against God and nature, and deserve the punishment and affliction of famine: especially in such a climate as Jamaica, that is subject to such violent vicissitudes, such perishing heats, and destructive storms, we should be ever watchfully provident, to have a good scope of ground provisions planted in patches, to become ripe in succession ; such as cocoas, yams, cassava, and sweet potatoes ; the method of planting which, I shall describe in another place. The space of ground they cover, is trifling in comparison to the cane and grass cultivation on an estate. Three acres of ground provisions, kept moderately clean, after they are established, will give an ample supply, even for years, for the support of the resident white people on an estate. The cocos, cassava, and sweet potatoes, will ratoon for two or three years; the negro yams are a yearly crop, but the white yams will last in the ground for several years. A gang of thirty negroes will clean and mold three acres of ground provisions in little more than a day, and all they require is two good cleanings in a year, yielding by this trivial attention a supply of provisions, in case of storms, which will in a great measure bid defiance to their effects. Other provision grounds, which will not bear storms, should likewise have their share of notice, that is, the plan-

tain walk, or walks, which an estate is generally provided with, but if suffered to go to
ruin, will produce very little. The plantain walk will stand very dry weather,
though not a storm, so that in case the
ground provisions should be backward, on
account of dry weather, the plantain walk
will furnish the table with excellent nutritious food, generally esteemed by the
negro epicure, as his best vegetable diet;
and the white man in Jamaica terms it
bread-kind, or a good substitute for it. I
have known some overseers, after receiving
the charge of an estate, where abundance
of all kinds of provisions were already
planted and established to their hand, so
shamefully neglect them, that in about
eighteen months, they had scarcely any
provisions to gather in and eat. They then
resorted to the expedient of bartering to
procure some, and had courage (or rather
audacity) enough to ask their employer, to
purchase flour for them, to supply their
table, instead of what they wantonly and

improvidently neglected. Let me here strenuously impress the necessity of paying impartial regard to this most essential requisite, to avoid the sad consequences of famine, by such fatal short-sighted conduct.

To conclude this part of my work, I must observe, and lay it down as a rule, that after the copper-wood, staves, &c. are brought home to the works, the mules should have rest for a fortnight, with good feeding, and attention paid to their wounds if they have any, before crop begins.

CHAP. VI.

CUTTING OF SUGAR-CANES AND MANUFACTURING
SUGAR.

HAVING (with as much precision as I presume to think attainable) brought the theory
and practice of plantership in general into
a system, so far as relates to the management of a sugar estate, down to the period
in which the sugar cane is deemed fit to
cut, in order to be manufactured into sugar;
I will now venture progressively to go on
in describing the method of cutting the
cane, and transforming its juices into sugar
and rum. The necessary apparatus for the
working cattle and mules being ready to
tackle them with; the mill in proper order
to bear stress of work, well secured and
clean; the gutterings, coppers, and stills
perfect, scoured, clean, and fit for the reception of ripe uncontaminated cane liquor;
the curing-house cleaned out, the mill-yard

and works cleared from weeds, grass, and all
description of filth and rubbish ; the water-
trenches about the mill-yard free from all impediment, so that no water will overrun it, and
the mill-yard swept, ready for canes to be deposited there; the mill-dam made up, to collect and supply an adequate force of water
for the wheel; and the leading water gutter
from the dam, perfectly cleansed from sand,
stones, and rubbish, so that the water will
have an uninterrupted course to flow to the
wheel, without waste or overflowing ; those
requisites being completed, and the great
gang furnished with efficient sharp bills and
knives, they should be set in to cut canes
(Providence giving favourable weather) on
the piece, either plant or ratoon, which first
becomes ripe for sugar making. A book
of cultivation, describing the course to be
taken, should always be kept by every good
and wise planter on an estate. The great
gang should be formed into two divisions ;
one placed in the rear of the other. The
front division as cutters, the rear as tiers of

the canes. The tiers, or rear division, to be only one-third as numerous as the front, or cutters; because the tiers will bundle the canes up faster, than the cutters can supply the proper number. A gang of forty negroes will compose thirty cutters, and ten tiers of canes. The gang should be set in on a fair breadth, generally at the farthest end of the cane piece first, as it may suit best, to prevent the ratoon sprouts as they come up being injured or eaten by the stock, when carrying canes or tops off the field, or as that part of the piece may come ripe first. Each negro in the front division should have a cane-row assigned him, and they should be made to cut their rows, or work in as regular a manner as possible. The rear division, or tiers, must keep regular pace or time with the cutters. The driver and book-keeper should have a strict watch over them, to prevent any sour, rat-eaten, or sucker canes being tied up, to be sent to the mill for sugar boilings. These will make bad sugar, plague and distract

21

the negro boilers, or contaminate several hundred gallons of liquor. The cutters should junk the canes, to a convenient length, for the mules to carry them. They must not be too long, lest they pierce the mules necks ; nor too short, to make an awkward load for the mule boys to place on the beasts.

It may sometimes happen, that plant tops are wanted at this period, or shortly after, to plant out, or supply a cane piece with. The cutters should in that case cut plant tops, as they proceed in cutting the sugar canes, making separate rows of them, and the tiers keep pace in tying them up accordingly; throwing them in distinct heaps from the sugar canes. It would be very advisable in every overseer to study the nature of the ground the cane grows upon, which he orders to be cut to make sugar from, whether it is rich or poor, light or heavy soil ; and the management such canes have progressively received to their maturity ; thus he may best direct the driver

how to top the canes, whether long or short.
Much depends on such skilful circumspection in making good sugar, with little boiling or fuel, and preventing subsequent
trouble, in correcting the bad effects of
spurious liquor being mixed with the ripe,
genuine cane juice ; which a little previous
and experienced forethought might have
obviated.   In rank, moist, heavy soils, the
sugar cane, when ordered to be cut for
making sugar, should be long topped ; that
is, the plant top should be left long, as the
juices in the four or five joints towards the
summit of the canes is poor, unmatured,
and difficult to granulate in the boiling ;
taking an overplus of lime, not only to prevent it turning sour soon, but to separate
the essential salt of the liquor from its
spumy muscous parts.   If this is not attended to, the sugar will be soft grained,
dingy coloured, take a great deal of boiling,
and expend much fuel in the operation.
Canes which are, or lately have been arrowed, suckered round the top joints, or

suffered to run foul by not being brushed,
require to be long topped; but canes which
have grown from poor, dry, light soils, are
not arrowed or suckered, and which have
been kept even moderately clean from trash,
do not need such long toppings. Two
joints in that case left on the top will be
enough, as the juices in canes from such
lands sooner become ripe, are richer, take
less time, boiling, and fuel, to convert them
into good sugar.

The borer, an insect of a subtle, pernicious nature, often infests both half grown
and ripe sugar canes. It is a dreadful malady
in this plant, which hardly any experiment
can get the better of. It rages like a plague;
its pestilence is insinuating and contagious,
and makes horrid havoc for some time, and
when not expected, it suddenly ceases its
ravages and disappears. This insect is so
diminutive, as scarcely to be perceivable.
It makes its way into the cane by a small
puncture through the rind, which quickly
(by its venom) appears black, communi-

cates its poison to the circulating juices of
the cane, and soon renders it an unfit substance to be converted into sugar ; at least
that part of the plant, and its vicinity, that
happens to be the seat of this destroyer.
The pith of the cane, by its active putrefying nature, assumes a light crimson color,
soon dries up, acquires a fœtid sour smell,
and all the part so affected, must be cut and
thrown away.   If it is mixed with the un-
injured, untainted part of the cane, which
is proposed to be made into sugar, the juices
will be perverted, and all the art and skill
of the manufacturer cannot turn it into
sugar.   I generally found that this pestilence attacked the canes which happened
to be foul, had a quantity of rat-eaten canes
lying in the cane-pieces through the rows,
and close to the stocks or roots of the canes ;
which in the middle of the year, when the
sun is unusually powerful, and after showery weather, caused quick and intense cor-
ruptive fermentation in such foul ingredients ; these generate such noxious insects,

and bring this curse upon the canes. As in the community of mankind, and the animal creation, cleanliness is not only conducive to health, but the principal means of prolonging life, and preventing distempers ; so it is likewise with the vegetable world ; all means to promote and secure their welfare by cleanliness (the best and most salutary of doctors) should be attended to by those who have the care of them, and to remove from them and their vicinity, whatever may either be of no use, or injurious to them. For this reason, whenever a cane-piece is cut down, I would put in a gang of negroes carefully to pick and tie up all the rat-eaten sour canes, have them brought to the mill-yard, heaped up there by themselves, and when no sugar is making, have them put through the mill. Any juice which may come from them may be sent to the distilling-house, and the trash carried to the trash-house for fuel. This would answer a double good purpose, in cleaning the cane-pieces from such an

annoyance, and adding to the stock of fuel.
Immediately after these pernicious canes
are removed from the cane-piece, I would
set the gang in to turn the trash off the
cane stocks ; give them air, clear them from
weeds or grass ; and if the sprouts are any
height, give them some mould. I am not
an advocate for burning off cane-pieces,
especially in dry weather parishes ; as the
surface of the ground, the trash, and other
light vegetable substances, which soon
make, or are converted by the weather,
into some kind of light manure, and keep
the ground and cane stocks cool, may be
injured by so doing. But when the borer,
or even many rats invade a cane-piece, I
would set aside this consideration, and resort to the alternative of burning off the
cane-pieces so infected, after I had picked
and carried off the ground the refuse rat-
eaten canes ; steady attention to cleanliness
in cane-pieces, keeping the trenches free
and open, no stagnated, putrefying water,
or pervading sour substances suffered to

remain on them, the rat-eaten refuse canes removed from them, the trash turned without delay, and lastly, (but perhaps not the least remedy,) that of burning off, after the cane pieces have been cut down, is the only feasible cure, that I think can be devised, with a hope of success, to annihilate and counteract the workings, of this cruel scourge the borer, from and among sugar cane pieces.

At such a period as crop time, the never-to-be-forgotten cure of the stock, should be a leading consideration with the manager ; some weakly people may be put in to follow the gang, which is cutting and tying sugar canes ; to tie up long tops for the cattle and mule pens ; that is, tops with the leaf attached to them, which are lopped from the spungy end of the plant tops ; they should always be gathered fresh and green, previous to being tainted with sourness. The stock would relish, eat heartily of them, and not be bloated or attacked with the gripes by feeding on them. They are strengthening enlivening provender, for both cattle and mules; greatly assist in furnishing the pens, both for day and night feeding, and spare the consumption of guinea grass, which would otherwise take place. I do not approve of turning stock into a cane-piece to feed, (which is very often done) as the negroes cut down the canes, as they tread, with a heavy impression, the ground too much; break and crush the projecting, and peeping cane-roots, which may bring on them ruin; and the feet of the stock are liable to be cut by the sharp ends of some of the cane stocks. However, in wet weather, they should never be turned to feed on a cane piece, except one thrown up, and it matters not, in what kind of weather, they are turned into such an one to eat it down. It should be a necessary routine of duty in every overseer on a sugar estate, to regulate the number of cattle and mules for work daily, or by spells or teams in crop time; either in carrying canes to the mill, turning a cattle mill, or drawing sugar to the wharf, and carrying provender to the pens. The mules and cattle which work one day, should not (if it can be avoided) be worked the next day. Sometimes it will be expedient to bring home canes to the mill, by both cattle and mules. Light carts should be had for the cattle, and plenty of good tackling for the mules. On an estate where bottoms in cane cultivation, lie contiguous to hilly land which is to be cut, fewer mules will answer. They can carry the canes, from the hilly distant land to the bottom, where the carts may be loaded, and thence by the cattle, drawn in carts to the mill. Where bottoms alone are to be cut, no mules need be employed in carrying canes to the mill; the cattle and carts will perform all that service. But when an estate is mostly hilly, or no roads fit for carts to be brought to act, the entire of the cane carriage must be done by the mules. It then requires an additional number of mules, according to the distance of the cane-piece from the works, but generally, I found, upon an average of distant and near cane-pieces, that fifteen or eighteen mules, in moderate weather, when the roads are tolerably good, were adequate to keep the mill supplied with canes, to produce fifteen hogsheads of sugar weekly. This quantity, one set of coppers, comprising two syphons and four low coppers, were equal to turn out weekly, to be potted in the curing-house. It then comes to bear, that if fifteen mules can keep the mill in canes, to make this return weekly, and that those fifteen mules can be spelled by fifteen others, at a stated time in the day, making thirty mules for one days' work, to carry canes to the mill; and that thirty more mules are in readiness, to work the following day for the same purpose, that the mules which work one day, need not be imposed upon to work the next; and that sixty mules will perform the requisite cane carriage to the mill, with ease, safety, and satisfaction, if well fed and tackled:

twelve more mules will be equal to do the
other by-work which may be wanted;
making a total of seventy-two effective
mules, to take off a crop of two hundred
and fifty hogsheads of sugar annually; with
other incidental work, which can be brought
in when the mill is not in use. On an estate,
where bottoms principally support the cane
cultivation, except a few patches of gentle
acclivity, few mules will be required, as
carts drawn by cattle, can always approach
close to the place where the negroes may
be cutting canes.   It then will require a
greater number of working cattle, to answer
the purposes of cane carriage, bringing
provender to the pens, mill-work, and drawing the produce to the barquadier. Cattle,
in the long run, turn out more profitable
than mules.   They make more manure,
and when condemned as unfit for work,
they can be fattened for the butcher, and
sold for little less than what they cost;
whereas, when a mule becomes infirm, or
unserviceable by age, disease, or accident,

it is a dead loss. No one will purchase such a live, worthless trunk. He prowls about the property, breaking fences, trespassing on some enticing provisions, which he instinctively finds out, is a mischievous freebooter, and devours what the effective stock should feed upon. A poll-tax is assessed for him equal to a prime beast, and I believe he is suffered to exist, chiefly on account, that his name should not be included in the decrease list, or bills of mortality for an estate ; than which nothing is more dreaded by an overseer, as his character for good or bad plantership, is in a great measure estimated, by this recorded criterion.

It is at this crisis of plantership, after a cane-piece has lately been cut down, when nature swells and animates the cane-root, when the earth makes it bud forth new shoots, that the fostering care of the planter is required, to administer to its succour and support. The trash which has fallen from the canes, (as they have been cut down,) lies promiscuously with dry cane tops, sour rat-eaten canes, and other rubbish, on the imbedded cane-roots, which naturally call to be relieved from its incumbent burthen, and silently implore the benefit of a free expanse of air, that they may quicken in their growth. The overseer should then, as soon as the cane-piece is cut down, (except he intends to dig and replant the piece,) set in the great gang to turn this load of trash off the cane-roots, place it regularly in the spaces between them, and clean the trenches out, to prevent water from lodging. It would be rather soon then to dig any mold to give them. They only want and demand freedom and air, to cherish and make them start from their stocks. This turning of trash off them, is transient work, and soon performed. But in about two or three weeks afterwards, or when the sprouts are six or eight inches high, it will be necessary to put a gang in, to give them a plentiful molding, in order to cover their roots and

feed their stems; and likewise to hoe off or
pull up any grass or weeds which may grow in
the spaces, or between the sprouts, and clean
the intervals and margins, from bushes or
weeds. This work comes in the line of method, which every overseer should prescribe
to himself, in the management of a sugar
estate, which if once lost sight of, will endanger and confound the whole system,
making it both vague, deranged, and ultimately unproductive.

Cleanliness should ever be strictly observed in every department connected with
sugar boiling, as it chiefly promotes the
manufacture of sparkling, strong-grained,
fair-coloured, marketable Muscovada sugar.
Slimy sourness will be prevented or coun-
teracted by frequent cleansing and washing
the mill-cases, mill-beds, liquor-gutters, receivers, and syphons; the leads of the low
coppers, strainers, ladles, and skimmers;
no trash should be allowed to remain long
in the mill-house, nor spumy pith in the
mill-bed or gutterings, which often gathers

imperceptibly there, soon ferments, and diffuses, by its slimy sourness, rank pestilence to the cane liquor, as it descends to the boiling house. A perforated plate of copper or lead should be soldered across the liquor-gutter, a few feet from the mill-bed, to prevent particles of trash or slime from running with the liquor; but this is so often neglected, that sometimes the receiver in the boiling house gets partly filled up with this poisonous stuff. A negro boy or girl should always be stationed in the mill-house, to keep this plate or strainer clear; and likewise the mill-bed and gutter free from trash. It has been a custom to keep a barrel in the mill-house to throw this vile stuff into, but I condemn such a practice, and would have a vessel, stationed outside the mill-house, at some distance to receive it. The best vinegar obtained on an estate is from the drainings of this barrel, which plainly evinces its quickness in becoming a strong acid. The smell of it is wafted to the adjacent cane liquor, its contagion spreads through it, the opposition of its nature is soon perceptible, more time is required to counteract its effects, and often to preserve the rest of the liquor in the boiling house from being contaminated, it is turned down to the distilling house, to be made something of there. I should always prefer this latter mode, instead of striving to restore, and make sugar of it ; for it never can be brought to concoct, or granulate, when far gone in acidity. It should be mixed with other good purified liquor in the low coppers for so doing, and all will at last be converted into a dull, dark, heavy, inspissated, and ungranulated substance, totally unmarketable. After spending much fuel in the boiling of it, the work will be at a stand for some time, as the coppers must be well cleaned from impurities before any more cane liquor should be allowed to go into them.

An ostensible slave officer, called the boatswain of the mill, should always be put in commission, whenever the mill is at

work, to guarantee, by his presence and
experience, the safety of it; to keep the
mill-gang to their work; to see that the
stokeholes are supplied with dry trash, the
green trash taken from the mill to the trash
houses, and well packed there; the mill
furnished with canes by the cane carriers;
the mill well braced; the cogs and gudgeons greased; the mill-bed, cases, and
gutterings, well washed three or four times
a day, and plenty of water to turn the
water wheel. This officer should be a carpenter, who understands the formation of
the mill machinery, can easily detect any
fault or tameness in its members, and quickly find a remedy for the defect. The mill-
gang and inferior boilers are chosen by the
head driver, from among the great and
second gangs, and are composed of such
effective people as he may deem proper for
the work. The head boiler should always
be an experienced negro in such work.
He is chosen by the overseer, to follow his
directions, and to conduct the critical busi-

ness. Sometimes there are two ostensible boilers, to spell and relieve one another ; but this breeds envy and strife between these jealous-headed people, and often confusion is produced throughout the work. It would be better, therefore, to have only one principal boiler, make him the responsible person in the boiling house, and when he is obliged to be spelled, for the purpose of natural rest, he should leave his injunctions to a judicious negro, whom he and the overseer can trust and put confidence in, to carry on the work in the boiling house till he returns. Many of the junior boilers are proud and emulous of such an undertaking, and often turn out excellent manufacturers of sugar. It will be well for the overseer not to chide or check the head boiler much, except a glaring fault occurs in him ; he may become dispirited, diffident, and careless, by so doing. It often occurs, that this man has a very general knowlege of the method of making good sugar, from almost every cane-piece on the estate, is conversant with the soils, the management the canes have received, and when the overseer may be in a dilemma, knows how to proceed to correct some perverseness in the cane liquor. This useful slave may, by his ready experience, explain the cause, and apply a remedy to prevent its bad consequences. The head boiler and boatswain of the mill are the leading, ostensible, and confidential persons about the works in crop time, while sugar is manufacturing.

As the cane liquor descends from the mill-house to the boiling-house, to be there collected in a receiver, there should be a coarse strainer made appendant to the end of the liquor gutter, and overhanging the receiver, to prevent the particles of trash and pithy slime substances from falling into the receiver, and mingling with the liquor in it. This strainer should be frequently washed, not only with pure water, but with thin lime-water, as it is very apt to gather and accumulate sour excrescences, to the detriment of any cane liquor which may run through it. I would have no more than one receiver, containing just as much as one of the syphons, because having a great quantity of liquor in two receivers, besides the syphons being full, the liquor by remaining in those vessels long, before it is drawn off to the low coppers, to be boiled and purified by skimming, becomes sour, or takes an overcharge of time to preserve it fresh. The slow heat which operates round the region of the syphons, and which is, in some measure, imparted to the receivers, soon causes the dreaded acidity in the liquor ; and before the low coppers, by rapid boiling, can cause sufficient evaporation, to admit of more liquor being drawn down from the syphons into them, the liquor in the receivers assumes an acid, slimy nature. I would, therefore, never have more liquor in the boiling-house than what will keep the low coppers constantly at work. If possible, they should never run short of liquor, except when boiling off on Saturday night, when plenty of water should be thrown into them, in order to have them cool, to be ready for scouring. By stopping fire, in order to restore, or regain a proper quantity of liquor to those coppers, a good week's work is interrupted, labour is lost, as some of the workmen remain idle; fuel is wasted; the coppers are apt to scale and burn; whatever liquor remains in them becomes rancid, cloudy, and skips indifferent sugar. Some overseers are passionately fond of night work, and send the mill-gang and boilers to the field during the greater part of the day, on purpose to push on some favorite field work, which, in my opinion, is a bad plan, if it can be avoided. There may be an excuse for such stoppages when they have not stock to supply the mill with canes, when there is a dearth of water, or the mill is not powerful enough to supply the coppers with liquor; otherwise, I presume to think, such schemes are untoward; have nothing but shallow policy to justify them, and are designed to cover some mistake or error that has crept in, and which they do not wish to be discovered.

The lime to temper the cane liquor, should be of the purest and strongest kind, devoid of stones, and in a pulverized state. The fairest way to apportion the quantity of it, to the specified number of gallons of cane liquor, the receiver can contain, is by weight. So many ounces of lime, to what the receiver may hold of cane liquor. Yet there is a nicety to be observed in so doing, and the experienced boiler must bring his best talents to bear on this point. The event of good or bad sugar being manufactured, greatly depends on tempering the liquor with lime precisely. Plant canes generally take more lime than ratoons, to cause the juices to granulate. When the canes grow in rank, heavy, moist soils, the liquor will take more lime, than when they grow in poor, light, dry soils. Well trashed, clean, unarrowed canes, take less lime than canes which have been kept in heavy trash

and weeds, and are top-heavy from arrowing. All this must be taken into the account,
and a nice circumspection used (knowing
these different circumstances) in tempering
the cane liquor with lime. I have seen
some cane liquor, take more than a pound
of strong lime to three hundred gallons,
and other liquor, not more than two ounces
of lime to the same quantity. This is a
problem which requires to be solved, to
make the efficacy of the operation certain.
The only way to elucidate it, to make it
perceptible to the mind, so as to form it
into a fixed practice, is to pay attention to
the before-mentioned remarks. There are
two motives for mixing strong lime with
cane liquor; one is, to prevent and super-
sede the souring of the liquor, the other to
cause it to granulate, (after it is purified by
boiling and skimming,) and separate its
essential salts from its spumy, mucous substances. The addition of lime is as a pre-
server; it does not clarify the cane liquor,
nor give a fair color to the sugar. On the

contrary, a large quantity of lime mingled
with the cane liquor, gives a dark blackish
hue to the liquor, makes the grain of the
sugar too large and hard, liable to crack
and scale in the coolers, and sink in the
hogshead when potted, shewing a scaly,
porous, light body, and gives a reddish
brown color to the sugar, not suitable to a
good market, and therefore not profitable.
I would strongly recommend, when cane
liquor is to be tempered with lime, that
only one half of what is requisite to be
given it, should be mixed in the receiver
with the liquor, purely to keep it from
turning sour; and after the liquor gets
partly clarified in the syphons, is drawn
from them into the low coppers, well purified and boiled there; passed to the tache,
and boiled there to a proper consistency,
then add to it, the remainder of what lime
may be necessary.   This the head boiler
will soon know, by dipping the ladle or
skimmer into the tache, drawing it out,
and observing how the syrup or sugar runs

off it, in seedy-like particles. If those seedy-like particles do not appear numerous, or the syrup is rather frothy, and runs off the ladle quick, he may add the remainder of the lime, or such part of it as will produce the proper appearances on the ladle. The lime to be added in the tache, (when the syrup comes to a proper consistency,) should be melted and diluted in some clarified liquor, and strained previous to being put into the syrup, that is boiling in the tache. Another method, which some people practice, to ascertain the crisis for skipping sugar, from the tache into the coolers, or to know when it comes to a proper granulating consistency is, to take a little of the syrup from off the ladle that is dipped into the tache between the finger and thumb, draw it to a thread, and if it breaks suddenly, when lengthened out, the sugar is sufficiently boiled. Another method is put in practice, to ascertain this critical point, which is, dashing a skimmer into the boiling syrup in the tache, drawing it out, and holding the skimmer up, resting its handle on the rail over the coppers, or keeping it in a position, so that the edge of the skimmer will drain off what syrup may be on it. When most of the syrup is run off, if there should be a thin membrane, or glossy web appear, pending from the edge of the skimmer; if this should break, and drop suddenly from the skimmer, without forming wiry threads, the syrup is sufficiently boiled, granulated, and becomes sugar; but if otherwise, it will take more boiling and perhaps more lime, before it is fit to be skipped into the coolers. But I prefer the former mode, as by it the double purpose is answered, of knowing when the sugar is properly boiled, and rightly tempered with lime; or the alkaline prevailing over any acidity, which may happen to have been in the liquor; besides the former mode of assay, is performed and acquired with more ease.

When the liquor is tempered with lime in the receiver, the vessel full, and a syphon ready to contain it, with a moderate fine strainer placed over the syphon, for the liquor to run through, it should be drawn off from the receiver into the syphon, but no grounds or filth should be allowed to follow it into that vessel. The damper belonging to the flue, which revolves round the syphon, should then be hauled up, to convey heat to it, till a thick scum gathers on the surface of the liquor, denoting the effects of the purifying heat. The syphon should not be allowed to boil, but as soon as this scum appears, the damper should be run down the whole of its groove to the bottom; and the negro stationed to mind the syphons, should take a skimmer, and gently take off the greater part of the scum, though not by any way to disturb the body of the liquor. I must here premise, that all skimmings and washings from the syphons and coppers, should be thrown into a gutter, and conveyed to the distilling-house. Nothing of this kind should be lost or wasted, as the distilling-house crop is much increased, by mixing it with the molasses, in preparing liquor in that department for fermentation, in order to produce rum. When the liquor is drawn off from the receiver, that vessel should be particularly well washed out with pure water, which will not take long in doing. It should then be filled again with liquor, and tempered with lime while filling. When it is full, and the other syphon ready clean, let the liquor be drawn from it to the second syphon, taking care to place the strainer as before on the syphon, for the liquor to run through it, and no mud or filth suffered to go from the receiver into that vessel. The damper belonging to that syphon should then be hauled up, till the operation of the heat through the liquor, causes a thick scum to form on its surface. Then let the damper down as before, take the scum gently off with a skimmer, without disturbing the body of the liquor. When the liquor in these two syphons is partially purified, by moderate heat pervading it, it is to be drawn down by a stop cock, from the first syphon to the low coppers, so far, as the convex part of the bottom of the syphon may be seen. Then stop the cock

for a short time, for the remainder of the
liquor for two or three gallons is muddy
and scummy stuff, (which descends with
the liquor, upon its surface, according as it
is drawn off to the low coppers,) and unfit
to be mixed with the rest of the liquor to
make sugar. When the cock is stopped
for a short time, (in order to prevent any
filth from running into the grand copper,)
it is to be again opened, and what filthy recrement is left in the syphon, should be
turned into the skimming gutter, which
leads to the distilling-house. This syphon
should then be well washed out with pure
water. All this should be run off the
syphon before the cock is stopped; when
well washed, the syphon should be again
filled with liquor from the receiver, and its
damper hauled up; then the liquor from
the second syphon, must be drawn off in
like manner to the low coppers, observing
to stop the cock for a short time. When
the bottom of the syphon appears, let the
dross be turned to the distilling-house, then
the vessel well washed with pure water as

before, the cock stopped, and filled again with liquor from the receiver. But so penurious are some overseers of the boiling-house liquor, that they will scarcely let a drop of the spumy stuff, from either receiver or syphons, be turned down the skimming gutter to the distilling-house; but all in a promiscuous filthy deluge, sent from the syphons to the low coppers; there to be boiled and compounded together, in order to swell the crop of sugar; the quality of the sugar being a minor object with them, and of little import what market it may meet with. So much they think their present good character depends, on a great number of hogsheads of sugar being made, that the quantity, and not the quality of the sugar is their sole object. The bulk of the distilling-house crop, is a matter of no moment to them; they think their employer will attach no blame to them, for any failure or deficiency there, because a subordinate white man superintends it. But who can make comparative good returns in the distilling-house, if the overseer denies

him the materials of sweets to do so; or filches
from him (because he is possessed of paramount authority,) the very dregs of the
boiling-house coppers, and boils the sugar
so hard and high, that little drainings of
molasses will issue from it. The resident
agent seldom visits the boiling-house. A
cooler of good sugar will be made now and
then by such an overseer, a sample of
which he dexterously sends to the agent,
who approves of it. He is lulled and
duped into an approbation of the abilities
and integrity of the overseer, and imagines
this sort of sugar prevails throughout the
crop; but it is only when the account sales
arrive from home, depicting the miserable
sales of the sugars, on account of their bad
quality, that this fraud will be clearly seen.
If this overseer happens to be a favourite
with the resident island agent, he only receives a rebuke, with injunctions to mend
his practice in this particular for the future.
I perfectly remember a most heinous trait
of this description, in a certain elderly over-

seer. The estate he had the management of, had two resident island agents, one a little more active and observant than the other. One of these gentlemen only visited the estate annually, the other somewhat oftener. However, the most active one was expected at the estate on a certain day. This overseer had been in the habit of making the negro stationed on the syphons, whenever he drew off the liquor from the receivers to them, send down every kind of filth with a mop, from that vessel into them; and when the scum appeared on the surface of the syphons, not to touch it, till the liquor was to be drawn off to the grand copper, and then with a mop, turning the liquor in the syphons, and mixing up all the spumy qualities with it, which had been thrown up by the purifying heat, to shove this adulterated body of liquor into the grand copper; where it was boiled up and only partially skimmed. But the day that the agent was to visit the estate, the overseer caused a transient revolution for the

16

better. All the old liquor was boiled off. Receivers, syphons, low coppers, and gutterings, were made perfectly clean, the boiling-house was full of liquor, and the syphons ordered not to be allowed to boil. The spume, when it arose, was taken off gently by the skimmer, and when the liquor was to be drawn off to the grand copper, the necessary contrived importance of the syphons was attended to. No mop was made use of, to confound the pure liquor and filthy spume together, and the liquor was gently drawn off by the regulating stop-cock to the grand copper, till such time as the bottom of the syphon appeared, and then the recrement remainder was turned to the skimming gutter. All these measures were taken in time; some good sugar was skipped on the top of what remained in the coolers; the agent made his appearance, was pleased with the state of things, and soon took his leave. The old system was gradually put in force again, with as much care as the spider returns to mend and re-weave his shattered web. However, this overseer only reigned there for another crop; his low cunning artifices were abolished by a voice from the mother country, complaining of the continual bad quality of the sugar; its sale was bad, and it was found expedient to send another manager to the estate. Overseers on sugar estates should beware lest they fall into such errors, for a momentary vain glory, or derogate from the true and honest system of making good sugar, with sound rich cane liquor. A strainer should be placed between the grand and second copper; a finer one between the second and third; and another still finer at the tache, on the skipping gutter. The two syphons of liquor having been drawn off to the low coppers, in the before-mentioned manner, strained from one copper to the other, and each copper, from the grand to the tache, having a due quantity of liquor (which these two syphons ought to be adequate to furnishing) the fire-maker or stoker should be called on to make a strong fire under the tache. The fire made in the furnace communicates, with amazing force and rapidity, by draft flues collateral to the chimney to the other coppers and syphons. According as the liquor slowly works up, by the force of the fire, the head boiler being at the tache, and the other three low coppers, having a negro each attending them, should be cleanly skimmed from any foulness or scum that is thrown up to the surface; such skimming may be turned into the gutter leading to the distilling house. This requires constant attention, according as the liquor evaporates in the tache, it is to be filled from the liquor in the adjoining copper; the adjoining copper to be replenished from the second copper, and the second copper from the grand; straining the liquor from one copper into the other, but taking care not to empty the coppers so far that they may burn, but keep their sides cool with liquor, till beyond the power of the fire to injure them. Nearly all the liquor should be passed from the grand copper to the other low coppers (to which it is tributary) before fresh liquor is drawn into it from the syphon. As what has been already boiled in the grand copper has been in a great measure clarified by skimming, it would be wrong to add impure to pure liquor. The grand copper being again filled from the first syphon, which has been partially purified by heat, the syphon, as before directed, should be well washed out, again filled from the receiver, and the constant routine of skimming the low coppers persisted in. The fire having penetrated through every aperture of the furnace and flues, by being constantly fed with the strongest dry mill-trash, its power becomes so great that the surface or leads, which unite the low coppers, appears a sheet of foaming liquor. Keeping up such a constant good fire (if the liquor is pure and good) will be a principal means of making fair-coloured crystaline sugar. The overseer, head-boiler, or book-keeper, should pay implicit attention to this main point. The liquor in the coppers being thus roused into foaming action, by strong incessant fire, the evaporation is great, and every opportunity is afforded to take away from it any dirty particles by the skimmer. At last, the eye is gratified by the liquor assuming a transparent appearance, of a colour resembling Madeira wine: a sure and happy symptom of good sugar being made. In this manner the low coppers should be replenished, the liquor strained and skimmed, till evaporation condenses the liquor in the tache into syrup, and that qualifying reducing vessel is full to the rim. Then let the head-boiler have ready lime diluted in pure cane liquor, which should be strained; and having dipped a ladle in the syrup to try its consistency and state of granulation, if it is found to want any more lime tempering, he can add what may be necessary to produce the desired effect; taking care to throw in a little liquor from the adjoining copper, so as not to let the syrup boil too

hard or high. In about a couple of minutes
after this operation the syrup becomes sugar,
and fit to skip into the cooler. This is ascertained by the head boiler ordering the
stoker to stop the fire, and open the cooling gate at the tache end of the furnace, so
that the coppers will not burn while the
sugar is skipping into the cooler. This being done, and the grating bars in the furnace cleared of ashes and cinders, the liquor
from the third copper passed into the empty
tache, that from the second to the third copper, and from the grand to the second copper, and the liquor from the syphon that is
longest purified drawn into the grand copper; then fire is again to be made under
the tache, and the same routine of steady
strong boiling kept up, straining, skimming,
and passing the liquor from one copper to
the other, as waste by evaporation demands,
till the liquor in the tache is again condensed into syrup, sugar again produced,
and fit to be skipped into the cooler. In
this manner the work is to go on in the

boiling-house, with little interruption, till the coolers are nearly full, or sufficient in them, to pot one or two hogsheads of sugar. As soon as the sugar is skipped into a cooler, the head boiler is to take a stirring stick, a long stout rod, made flat for eighteen inches at one end, and three inches broad where it is flat, and with this stick work the sugar to and fro in the cooler to mix it well, then gently pass the end of the stick over the surface of it, till it is streaked by so doing all over. Then let it rest for about twenty minutes, till it forms a crust on the surface, at which time, before it is too cool, take the turn stick again and pass it over the surface gently to break the crust, and give a crystalised variegated appearance to the sugar in the cooler. This method is to be observed with every skip of sugar that is passed into the coolers.

When a sufficiency of sugar is collected in the coolers to pot one or two hogsheads of sugar, and that it is so cool that the finger may remain in it, then let a couple of seasoned well-hooped hogsheads be placed in the curing-house, on the rangers, and made to stand level, so that the sugar will lie even in the cask, and, when full, have an even level surface. The hogsheads should be as firm on the rangers as possible, for, if unsteady, they may give way and cause unpleasant accidents to the negroes or sugar. The head-boiler or book-keeper should superintend this business; a good deal depends on the manner in which the sugar is potted, to make it stand the cask well, not sink much, cure properly, and turn out when weighed the proper quantity, conformable to the size of the hogshead. A couple of negroes, with middle-sized pails to carry the sugar, another with a strong spade and shovel to dig it in the coolers and fill the pails, will be sufficient for this work; enough of invalids can be found for this purpose, without drafting away able people from other work to do it. The sugar, if hard in the coolers, should be chopped, but not so much as to injure the grain of it. No large lumps should be sent to the hogshead; they should be broke moderately small, and the whole mixed well in filling the pails. The hogsheads, if tight in the bottom or very close, should have four or five small augur holes bored in them, and some plantain stalks run through the holes and drawn perpendicularly to the top of the cask, where a cross stick should be placed, likewise bored in various directions, and the stalks run through it, in order to let the molasses drain off from the sugar in the hogshead. The negroes should throw the sugar from the pails into the hogsheads all round, and in the middle, and as they gradually fill them, have a stick to make the sugar lie close and dense; and even in the casks. If the sugar is not too soft or hot, it will preserve its position in the casks; not sink or break into inequalities, and be firm in an hour or two. If it is crisp hard-grained sugar, it will cure in three weeks and be fit for shipping. I shall here conclude my observations and instructions on the management of sugar-canes, and manufacturing their juices into sugar, by stating, that though much depends on the care and management the cane receives in the field, to be successful in making bright strong-grained fair-coloured sugar, together with a due and satisfactory quantity of it, according to the means which can be brought into action for that purpose : yet this prudent judicious management may be in a great measure defeated by wilful neglect, ignorance, inattention to cleanliness, bad fires, or clandestine plotting in the boiling-house. So that though much has been done in the field towards prosperous results, the operations in the boiling-house should ever claim due attention, as there the long weary trial of patient skill is to stand a philosophical test ; for a happy or an unfortunate termination.

Once more I shall beg leave to introduce to the notice of my readers and the planters, the necessity and effect of manuring high for cane culture ; as the rearing of that plant causes great exhaustion to the land. And further, strongly to urge the

good consequences which follow from confining the cane cultivation to an improved
compact compass; and having all the cane-
pieces within that compass well manured
and taken care of, instead of stretching an
unbounded, scattered, ill-attended, and ill-
cultivated field of canes, that would be
accompanied and saddled with immense
labour, loss of stock, and indifferent returns of sugar. As an instance of this false
and destructive management, I shall here
beg leave to introduce a circumstance of
deplorable and unbounded extravagance in
a resident island agent ; or rather to point
out his partiality for waste and destruction,
his ostentatious prodigality, his tyranny in
prosecuting it, his deception in tampering
with and ensnaring his constituent into a
belief of the immense advantages to be derived from his ill-fraught, ill-digested, but
selfish scheme ; his bold, but weak, rash
undertaking ; his choice of overseers to pro-
secute his darling object. The fate of this
estate, and its entire capital, will be remembered for many years by those concerned

in it, and exist, as an indelible mark of folly and vanity.

In an angle of the parish of St. Mary's, in Jamaica, bearing nearly equidistant from Port Maria, Oracabessa, and Salt Gut, and about three miles from the borders of St. Thomas in the Vale, lies an estate, whose amphitheatre of hills embosoms, and nearly wraps up its compact, eligible, set of works. Its only public road, for the carriage of its produce to the wharf, winds by the course of a rivulet, which, after running in a serpentine direction, for about a mile and a quarter, joins the main road in a corner of the vale of Bagnals. The almost insular position in which it is placed, and sequestered station, the narrow winding road leading to it, the view of its durable works, which bursts suddenly on the sight, and its surrounding abrupt hills, which are cultivated with sugar-canes, in some places nearly to their summit, impress the stranger (nay even those acquainted with it) with the romantic sublimity of the place, and draw forth an expression of admiration, as to the persevering, arduous labour of man, in establishing it as a sugar estate. A most noble spectacle in the scenery here is, a stupendous white shelving rock, that with dreadful grandeur spreads to a considerable curve-like breadth, and upwards of one hundred feet in height, overhanging in projecting strata, from its immense base to its impending summit, which threatens momentary destruction to every thing underneath ; yet the hand of Providence, and natural adhesion of the massy strata of this frightful precipice, keeps it from tumbling to the bottom, and preserves unshaken its terrific aspect. In rainy weather, from the top of this rock, and out of a channel, (which time has formed, by the action of an impetuous stream from heavy rain,) descends a torrent of water, into a natural basin below. This cataract is seen from a great distance after heavy rain, and the foaming surge made from it, causes a roar, and mist to arise, which is truly striking and picturesque. Even in dry weather, when some springs cease to pour forth a

supply of water, the strata of this rock is
always seen to percolate, and drip down to
the receiving basin beneath, a gentle shower
of filtered, limpid water. From this basin
likewise, as well as from some neighbouring
springs, a large mill-dam is supplied, whose
water turns an extensive mill on the works
of this estate, and is also the parent of the
rivulet which flows at the base of its hills,
along the margin of its winding, barquadier
road, that at last loses itself in the Rio
Nuova river.

The resident island planting agent for this
estate had to deal with a constituent, who
though residing in the island, was wholly engrossed in mercantile speculations, a stranger
to the duties and qualifications of a planter,
and had his residence in the emporium of
the island. He confidently trusted to the
skill and unbiassed conduct of this (as
he thought) efficient planting agent. The
estate was a partnership concern for some
years, and one of the partners resided on it.
It was involved in a chancery suit. In the

year 1809, the partner who lived on the
estate was induced to leave it, and resign
his claims for a certain consideration to the
representative of the other partner, who
now became its possessor, although still
pending in chancery.   When the resident
partner (who was likewise the overseer of
the estate) was about leaving, it was found
necessary to send an overseer to obtain possession from him, and take charge of it.
These changes were anticipated for some
time.   Every thing was in a state of back-
wardness for commencing crop, though in
the month of March.   No copper-wood
was cut; half the pastures in ruin; no land
prepared or dug to put in spring plants; and
the coppers in a deranged state; but few,
and they all worn out, crippled Spanish
cattle on the estate, to make manure and
carry the produce to the wharf; immense
jobbing, which previously was spent on the
estate, was suspended or withheld; and it
was not till the month of May that the overseer who received possession, and took

charge, could get things in a state of forwardness to commence crop, and pre-pare ground for a spring plant, in order to ensure some satisfactory returns the ensu-ing crop. This overseer had the toil and anxiety of bringing such a property, from so low a condition, to comparative prospe-rity, and advantageous crops, without much jobbing, labour of hired mechanics, little loss of stock, none of effective or working slaves, and had an increase of slaves far beyond the decrease. A field of one hun-dred and sixty acres of promising clean canes, he had established, which the prying eye of a critical planter did not find fault with, but approved of, and had the works newly covered with shingles, procured from the woods of the estate. The mills and mill-houses were put into complete repair ; the boiling-house coppers newly hung, plenty of strong mill-trash brought into the trash-houses, to boil the sugar with ; cop-per-wood cut, and brought home to the works for the distilling-house ; the cane and

grass pieces cleaned; the cattle in comparative good order, the mules in excellent
condition, and every thing ready, in the
early part of the year 1811, to begin crop.
He had actually commenced an expectant,
promising crop, of near two hundred hogsheads of sugar; when the capricious, de-
signing, venal policy of the resident island
planting agent was put in practice, to supplant this overseer. He sent a well-feigned
story to his constituent, in order to get his
acquiescence for discharging him; thus
cruelly and barbarously taking from this man
his situation and his bread, and, if possible,
his fair, hard-earned reputation and character; and this without deigning to allege a
fault, or impute a crime to him, and throwing into the hands of the succeeding overseer an estate which was in excellent condition of cultivation, to build his reputation
(if possible) upon the spoils of that of his
predecessor. All this was done to answer
the purpose of a deep-laid scheme for his
own private advantage. But fate, though

slow, ordained it otherwise ; for the succeeding overseer was a novice of a planter, a
creature of the agent's will and pleasure,
though illiterate and ignorant; a person
whom he was resolved to promote, at the
risk of propriety and discretion, because a
secret combination of circumstances linked
him to this man, which he did not wish the
world to know of. I could here relate some
of those secret circumstances, but it would
break in upon my narrative of what happened on this estate, and the reasons that
induced this agent to dismiss the former
overseer; and fill his place with another.

When the storms of a western revolution
threw a number of refugees on the British
West India islands (some of whom were in
a destitute and naked condition) as adventurers, to seek an asylum and employment
there, the native hospitality of Jamaica received a number of those wanderers, and
without prejudice or partiality, or any dis-
parity, on account of what principles they
might profess, generously ranked them as

part of the island family, and employed and patronised them. In thus giving protection, and affording impartial opportunities to the deserving and the upright, they never reflected on the licentious views which might have been transplanted with those emigrants, or the dark but aspiring policy which has marked their gradual exaltation. They have now mastered the Floridas, and will soon have a powerful navy, both stationed in, and sweeping the gulfs of Florida and Mexico. We may look to Cuba and Jamaica for their next enterprising attempt; for they will seize an unguarded, favourable moment to attack and possess themselves of those valuable colonies. No matter in what corner of the world, any of those ramblers were, or are placed; a latitude has been given to their endeavours and enterprises, which has levelled almost every thing to their mental standard, and made them partake of the enjoyments and privileges, not only of British subjects, but of citizens of the world.

An estate in the parish of St. Thomas,
in the vale, (nearly two miles to the north-
east of Bogwalk, which verges on a branch
of the river Rio Cobre,) nurtured a middle
aged, but now nearly superannuated soldier,
of this western tribe of emigrants. He
there basked in the sunshine of prosperity for
some years, and patronage accumulated on
him, till his ambition made him vain, superstitious, and at last not very scrupulous how
this estate, and others under him, was attended to. Always aiming at personal
aggrandizement, he did not regard a little
infraction of justice. After a number of
years spent with various success, he ventured, in his advanced age, on the solacing
scheme of matrimony, and united himself
with a young Creole lady, not twenty-five
years old. In the natural course of time
she blessed him with so many children, as
to alarm him, and cause his brain to be
prolific in ways and means for their support, education, and future fortunes. Retrenchment was first adopted, his constitu-

ent's property was next assessed, parti-
tioned, and loaded with the incumbrance
of his family, both in point of house-room,
provisions, servants,, small stock, grass, and
corn. He then looked about with anxious,
unrelenting solicitude, for a situation suitable to his views, central in a great measure with some affluent, comfortable estate
that he was concerned for; and after various perambulations, and artful enquiries
about the benefits to be derived from such
a situation, and a specious display of what
great service his near residence would render to any estate he was employed to manage, he insidiously worked on his unwary
and credulous constituent, and was allowed
to remove to, and take up his quarters, in
the great house of a cultivated mountain,
adjoining the estate I have already described, as lying in an angle of the parish
of St. Mary's.   On this he displayed his
talent of extensive tillage, but brought poverty on the land, with no increase of crops.
This event, as I said before, was prefaced

with the discharge of the first overseer, because he imagined that man (as he often said, was of rather too independent a turn of mind for him) was not, or would not, be congenial to his schemes, and that contributions for his table would be but slowly and scantily sent in.

Although this estate had a number of excellent Negro carpenters upon it, and belonging to it, who were fully adequate to what work of that kind might be wanted, yet this agent, in the unrestrained latitude of his spirit, called in the assistance of a white carpenter, at a considerable expense, to erect a shoot, or strong wooden trough, in order to precipitate the canes from one part of his field to the bottom of the hill. This was a piece of rude mechanism, which the estate carpenters could accomplish, with no expense but their own labour. He not only heaped this expense on the property, but threw up and destroyed a valuable productive cane-piece of eleven acres, which lay contiguous to the works, to make room for,

and erect, this favourite, worthless trunk,
that in two years after was found of no use,
but rather an inconvenience; at the end of
this time its death-warrant was announced,
by pulling the tottering, rotten fabrick
down. But to proceed with this man's extravagance and waste of valuable land. On
the summit of the before-mentioned romantic rock, there lay a piece of ground, covered with valuable hard-wood timbers, and
such trees as were fit to split for the purpose
of hogshead staves and shingles, which
served for some years, and would have
lasted many more, as a resort for the estate
carpenters and coopers to procure those
requisite and necessary materials in, and
save the expense either of more laborious
carriage, or purchasing foreign lumber. A
track had been made some years before, for
mules in dry weather to proceed to it, to
carry home the lumber that had occasionally been split there; and although this
land was valuable as a repository of wood,
yet it was cold, clayey, and rather a spongy

soil, unfit and unprofitable for the cultivation of canes, a bad, distant, and a broken
up track or road, for carriage to and from
it; and what was of infinite injury to any
canes that were planted there, the entire
neighbourhood of this land was surrounded
to a great distance by ground which har-
boured an endless number of large rats.
To the extent of twelve to eighteen acres of
this wood, did he order his novice of an
overseer to have cut down, burned off, and
planted with canes, though even the plant
tops were obliged to be carried by mules
from a great distance to it, through a bad
road, over the stumps of trees, and through
tough, boggy holes. This land was separated by a high rocky hill from any of the
cane-pieces, or cultivated spots on the estate,
and unprotected by fences. Although land
of far better texture for the cultivation of
canes was on the estate, which was divested
of wood, and had not been much exhausted
by tillage, lay more commodious for carriage, had an exposed aspect to the sun,

13

(which this wood land had not) and much nearer the works than this, which some manure would have made adequate to produce three hogsheads of good sugar per acre, yet it was passed by unheeded. After bestowing the most arduous labour of the negroes and mules, cutting down, burning, and destroying this valuable wood, and striving to establish this land as a cane-piece, it yielded only two poor crops of bad sugar. The rats became masters of it, and it now lays a ruinous memorial of the folly and extravagance of this vain resident planting agent. In enumerating the almost endless, and I may say fruitless folly of this man, in the planting system, I should too much digress from my main subject. Suffice it to say, that he not only destroyed this wood land, by putting it in canes, which was thrown up shortly after, but he made this and two other successive overseers mutilate the estate, and its fine grazing pastures, and devote one-third of them, at a great distance from the works, to cane-culture: he

half-starving the cattle and mules, a great
part of whom perished by overwork,
and want of the plenty they were accustomed to. All this while, the old cane-
pieces of the estate were turned up, dug
and planted in canes, without intermission
or rest, and little manure, save one solitary
piece. Even the very negro grounds, that
had thrown up guinea grass, and which was
serviceable and convenient, for cutting as
provender, for the mule-stable, and cattle-
pens, was taken up in cane-pieces ; so that
an immense field of canes was spread out
in every direction. Such an extended
field necessarily required a great deal of
negro labour, to plant and preserve. A
jobbing gang belonging to a neighbouring
mountain, was by this agent's command,
called in to aid this work. Not less upon
an average than 40 effective slaves daily,
throughout the whole year, were constantly
employed on this estate, at this period,
which could not be less than an expense of

1700*l.* currency of Jamaica annually. But these projects were ill-judged and ill-executed, in a great measure frustrated by their overgrown weight, and after a lapse of three years, the bubbling scheme burst. The inutility and burthensome expense of it, was too apparent to the constituent; his selfish views became visible, and his dismissal proclaimed the wisdom of his employer; saved the estate from farther defalcation; and it is to be hoped, drew the sphere of cane cultivation on it, into a narrower, more healthful, and more profitable condition. With all this immense jobbing, of 1700*l.* per annum, they could not preserve this large field of canes clean; they yielded poorly, because they were planted without judgment as to season, without manure, with imperfect tops, badly fenced, rammed into the cane holes in a careless way; cut in all weathers, to enforce the mill to grind them in time; and trash turned too late on the pieces, without supplies. This agent never made up an annual crop, during the three years I before mentioned, of 200 hogsheads of sugar, which the estate could have produced from a field, of 60 or 80 acres less of cane culture, in far better order, and a saving of 800*l.* or 900*l.* a year.

But though vanity and pride were the principal features in this agent's character, he never could divest himself of an inclination to support himself and family, at the expense of his constituents. No diminution of his per centage occurred, in the account sales of the produce; or reduction for the maintenance of himself and family. This was amassed with rigid parsimony. Though wealthy, he was ever needy. The spirit of avarice was never abandoned by him; and though detected in his meanness, his schemes exploded, himself and family banished from his former abode, he still, with matchless effrontery, sought for another asylum of this kind, and established himself on a property belonging to another of his constituents, who resided in England; and made changes of overseers on several plantations, to suit his convenience and interest.

CHAP. VII.

CLEANLINESS is a principal means to produce not only good, but large quantities of rum. For this purpose, every vessel, which is to be made use of in its manufacture, should be kept as clean, as the nature of things will admit. The skimming molasses, dunder, mixing, and fermenting cisterns, should all be well cleared out, preparatory to a rum crop. Even the tank should have its share of attention in that respect, the stills be brightly scoured inside, and the worms forced with water, to discharge any scum, mud, crustated stuff, or other matter, which may obstruct the free passage of the distilling liquor, or adulterate it. It is sometimes difficult in the commencement of a rum crop, (when the distilling-house is cold from disuse,) to bring on free, quick, and strong fermentation in the liquor, after it is compounded. It will be requisite, in order to promote this essential property, to bring into action something to assist its natural efforts. The fermenting cisterns should be well cleaned and dryed out, then filled with some green milltrash, which has not been much squeezed, that the fermentation arising from it may sweat and warm the cisterns. This trash should remain in them covered up, till they are wanted to be filled with liquor from the mixing cistern. The house should have a fire made in it so central, that the warmth of it will diffuse through it, and dispel the chilly cold dampness of the fermenting part of the house. The liquor should be compounded (the first round of the fermenting cisterns) rather lightly of heavy sweets, so as to induce by its volatile light quality, quick fermentation, and that in two or three rounds of the house, by gradually strengthening the liquor with strong heavy sweets, the standard of fermentation may be critically fixed, to answer the disposition of the house, in yielding good returns from the liquor. The house being thus regulated, nòthing is wanted but cleanliness, economy, and a due share of sweets from the boiling and curing-houses, to give an adequate return of rum, to the number of hogsheads of sugar made ; that is, one half the number of puncheons of rum, to the number of hogsheads of sugar ; which if produced, is a criterion to judge by, that the distilling-house department has been managed with attention and ability. But a parsimonious conduct in the affairs of the boiling-house, may quash an impartial crop in the distilling-house, and render both crops poor and unproductive.

No old dunder (or wash from the liquor still) from a former crop, should be made use of to mix with liquor of a new crop. This pernicious stale stuff, should be thrown away, and the dunder cistern left empty, to receive the new dunder from the first

liquor that is distilled; good fresh clarified
dunder is an ingredient, when given in
right proportion, which will enhance the
strength and fermentation of the liquor;
but if of a weak, muddy, and bitterly sour
nature, it will ruin all the other ingredients,
retard fermentation, make it work heavy,
take a long time to ripen, and give poor
returns. If the skimmings have not been
purged from spume and dross, and in a
manner clarified, previous to being mixed
with the liquor preparing for fermentation,
they will cause the liquor to be ropey, work
heavy, be slow in fermenting, vapid, and produce small returns; it then behoves a person superintending the distilling-house, to
be nice and exact as to good clean dunder
and skimmings.

Twelve per cent. sweets will be high
enough to set the liquor, in the beginning
of a distilling house crop; but when the
house is warm, with good strong fresh
dunder to make use of, and powerful fermentation pervading all the cisterns, I

would set the liquor as high as fourteen, or
fifteen per cent. of sweets. A measuring
rod is a useful implement, to ascertain the
quantity of each ingredient. This may be
put into the mixing cistern, with a per
centage scale scribed upon it, equal to the
depth of the mixing cistern ; so that, if the
mixing cistern holds twelve hundred gallons,
this rod should be marked with twelve cross
scribes, and the space between each scribe
divided by ten nicks ; each scribe denoting one hundred gallons, and each nick
ten gallons. The skimmings and dunder
having been in some measure clarified,
this twelve hundred gallon mixing cistern
I would proceed to fill with its compound-
ing ingredients, at fourteen per cent. sweets ;
that is, one hundred and sixty-eight gallons
of heavy sweets or molasses ; taking it for
granted, that every eighty gallons of good
skimmings, is equal to one gallon of molasses.   I would have turned into the cistern
four hundred and eight gallons of skimmings, which is equal to sixty gallons of

molasses. It then requires one hundred and eight gallons of molasses to make up the complement of sweets, which should be likewise thrown into the cistern. Having its proportion of sweets, and being filled up to five hundred and eighty-eight gallons, with skimmings and molasses; that is forty per cent. on the measuring rod of skimmings, and nine per cent. of molasses; the cistern now wants six hundred and twelve gallons of liquid to complete it. I would then add, for this purpose, four hundred and thirty-two gallons of dunder, and one hundred and eighty gallons of pure cold soft water; that is, thirty-six per cent. of dunder, and fifteen per cent. of water on the measuring rod. Total, twelve hundred gallons, being the contents of the mixing cistern, and mixed at the rate of fourteen per cent. of sweets. This liquor so composed, should be well stirred up and commixed, by a perforated broad board, placed to a well-fixed staff or handle. The liquor must then be left to rest, till it begins to ferment. The foul drossy head should be skimmed off it, and the fermenting cisterns being ready for its reception, it should be pumped, or allowed to run into them till they are full. In this manner should liquor be set in the height of crop. The liquor, when it is passed into the fermenting cisterns, must be kept clean by a skimmer; when no skimmings are to be had, the full quantum of molasses must be mixed at the rate of fourteen per cent.; that is, one hundred and sixty-eight gallons of molasses, with fifty per cent. of good pure dunder, or six hundred gallons of it; making together with the molasses, seven hundred and sixty-eight gallons. The cistern must be filled up with pure cold soft water, and will require four hundred and thirty-two gallons at the rate of thirty-six per cent. on the measuring rod, making a total of twelve hundred gallons, being the contents of the mixing cistern, comprising fourteen per cent. molasses, fifty per cent. dunder, and thirty-six per cent.

water. When beginning to ferment, it should be skimmed, and then pumped or passed to the fermenting cisterns. In this manner should the house be set round with liquor. When the fermentation subsides, or ceases in the fermenting cisterns, the liquor is ripe for distilling, which may be known by whitish bead-like particles, or small globules appearing on the surface of the liquor, or a thin white surface shewing itself on it. There should no time be lost in distilling it. It should be passed into the low wine still, the still well closed, and a strong fire put under it, till the low wine begins to run slowly from the worm. A moderate fire must be kept up, till the low wine is run off. According as the fermenting cisterns are emptied, they should be washed out with warm water, and filled immediately again, so that the fermenting spirit may be retained in them.

Good strong liquor should give a fifth from the still in low wines. Care should be taken to run off the low wines cool from the worm, and no liquor allowed (by the strong a fire being made under the still) to descend with it, through the worm. When there is enough low wine made to fill the rum still, or a couple of hundred gallons more, it should be passed into it, the still well closed up, and a quick strong fire made under it, till the rum begins to trickle from the worms. Then most of the fire should be damped or withdrawn, and the rum suffered to run cool from the worm. Good low wine will give one-third strong rum, or rather more. Plenty of water should be running into the tank, to keep the worms cool, and the surplus warm water, which rises to the surface of the tank, should be let off by a proper outlet. Neither the low wine or rum still should be filled higher than within six or eight inches of the rim of the still, that no accidents may take place by explosion, or any of the liquor or low wine from the stills, be suffered to pass down the worms, to contaminate the distilling liquor. The still heads and goose necks, must always be well closed and secured, whenever the stills are loaded with liquor or low wines, and a brisk strong fire made under the stills to bring them down, or make sufficient ebullition by boiling. After this, moderate fires will work off the distilling liquor. For the purpose of preserving good dunder to mix the liquor with, the entire spirit from the liquor still, (or rather low wine still,) should not be run off the still, but a few distill-house cans retained in the still, of the last runnings of the low wines, so that the dunder may be strong, and kept fresh and good, to mix in subsequent liquor, which it much enhances the strength and fermentation of. When running off, or distilling rum, a proof bottle should be kept, into which (by a phial containing a certain small quantity) a portion of the rum, from each distilling house rum can should be put; and when the rum so put into the proof bottle comes to the mean standard of twenty-two strong proof, by a London or Glasgow proof-bead, the remainder of the spirit, which comes from the still worm, should be thrown up into the low wine butt. Any more mixture of it in the rum, will both make it weak, and give it a bad taste and flavour.

Every week the stills should be completely scoured, and every time they are filled they should be well washed out with water. The dunder and skimming cisterns should be cleaned out every week or fortnight.

CHAP. VIII.

PLANTING AND CARE OF PROVISIONS.

IMPERFECT as this treatise may appear to
the nice discerning eye, to the critical observer on speculations of this kind, I do
not mean to arrogate to myself a vain presumption to enquire into deep researches,
but to give a plain unpolished system, as
to the theory and practice of managing a
sugar estate. I feel, in taking upon myself
such a task, that I should be guilty of a
gross error, and an unpardonable fault, if I
did not, in some measure, advert to the
prime rule of planting and taking care of
as much vegetable provisions as will suffice
to sustain the resident white people on such
a property. It is rigidly incumbent on
every overseer or manager not only to have
a knowledge of the method of planting

such provisions as are natural to the soil,
but to have plenty of them to resort to,
for the support of himself and the white
people living with him ; and for the benefit
of some of the slaves, who may happen to
be devoid of provision grounds in bearing.
A magazine or reserve of ground provisions
will always be of great service.   It is true,
that an annual supply of salt provisions,
with five or six barrels of flour, comes out
to almost every estate for the use of the
white people and slaves ; but this will go
but a little way indeed towards their maintenance.   It will be a mere mite to depend
upon.   Almost every article for human
subsistence must be raised by the planter
and slaves themselves.   The rearing and
care then of provisions is concomitant with
the principle of plantership ; for it would
be imperfect and lifeless without it.

Some managers have a hiding propensity
in the choice of a situation to plant provisions in ;   for which purpose they fix
upon one of the most remote corners of

the property, and care not how much valuable wood they destroy by so doing. At some future period this may be wanted for buildings and copper-wood, and save the expence and delay of purchasing a supply. They may imagine, perhaps, that provisions require prime virgin land to grow in, that they shall never come in contact with any other cultivation; that they will be censured by their employers, for bringing them in the scope of cane or grass tillage, or devoting time, labour, and attention to them. Most of these reasons I will venture here to refute. But at the same time, I beg leave to explain that I do not mean that either cane or grass cultivation should be injured, or cut up, by the introduction of provisions in their place, or that they should be planted close to the works, to the disparagement of other culture; that they may be close to the view, and under the eye of the overseer. With respect to new land being taken to plant provisions in, it is a waste to employ it so,

when many patches of good strong land
lie unoccupied in various sheltered corners
contiguous to the cane culture, capable of
throwing up strong luxuriant provisions.
These have been deemed, perhaps, too distant for the carriage of canes, too laborious
and tedious for the stock, to bring home to
supply the mill, and not wanted to convert
into pasture, as enough of grazing grounds
may have already been established for the
support of the stock. To this it may be
added, that by having provision grounds
for the white people close to the farthest
cane cultivation, any weeds that might
otherwise grow there, would be kept down,
and done away with, to the advantage of
the canes ; because the system of cleaning
the provisions so close to them would be
an essential safeguard, a benefit to the
canes. A watchman would be stationed
there, the subordinate white people would
often frequent the place, and have provisions cut and gathered; and the overseer's visits would be necessarily drawn to

such a convenient spot. By taking up land
of this kind in provisions, the new virgin
woodland would be preserved, the cane
culture would not be intruded on or injured, and labour would be saved. I have
no doubt but every reasonable employer
would be satisfied with the work, and
applaud the undertaking as a laudable one.

So much corn is usually grown through
the cane pieces, that seldom a separate
corn piece is wanted on an estate to afford
a supply. However, when the overseer is
forbidden to plant corn in the cane pieces,
he must have a separate piece of ground,
as an alternative to sow corn in for the subsistence of small stock; on which himself
and the other resident white people are to
live; for occasionally giving to the negro
children, convalescents, and weakly dis-
tempered mules and cattle. There should
in such cases be a piece of about ten acres
of tolerable good land set apart, contiguous
to the cane culture, and fenced in, where
liable to trespass from cattle, by a post and

rail fence, or a double ditch, penguined over. This piece of ground should be stocked up and hoed off; and if heavy in grass or weeds, cleared by burning when dry. I would have this ground regularly lined out, at four feet and a half distance between the corn-rows, which should be pegged off every two feet; then set in a gang of negroes with hoes to dig corn-holes, where the pegs have been put down, through the entire piece, which will be planting the corn four feet and a half space, by two feet distance. The corn-holes should not be more than five inches deep, and six inches square when opened. The mold must be hauled up to the edge of the holes, broke fine, and made a small bank of. The banks and holes through the piece must be parallel to each other. Then let each negro have a basket of good ripe manure, and put about half a handful in each corn-hole. When a favourable opportunity offers, I would set in the gang to plant it with good seed corn, which should be steeped in water for a few hours previous to planting. Let the gang be set in, a negro to each corn-row, each negro being provided with a small calabash or vessel to hold his quantity of seed corn, and make each of them drop four distinct corn seeds in every hole; not too close together in the hole, and then cover it up with the bank mold. If the weather is seasonable, it will in a week begin to shew itself above ground. When it is four or five inches high, the weeding gang should be set in to clean, weed, and mould it up. It will then grow rapidly, and will require another cleaning and moulding in four or five weeks' time. This crop of corn will be ripe for breaking or gathering in, (if favour-able weather,) in four months. The corn planted from March to October is generally the best and most productive. There is one advantage in having a separate corn piece from cane cultivation. It can be underplanted, in the spaces between the corn-rows, when the first corn is near three

months old, by digging corn-holes in the
middle of the four feet and a half spaces,
not exactly opposite the former corn-holes,
but at a mean distance between them.
These holes should likewise have a small
quantity of manure, and every time the
piece is planted, should have attention paid
to this direction, to keep up the strength
of the land; which will ensure good corn,
if the weather is favourable and the seed
good. By thus underplanting, there may
be two, if not three, crops of corn from
the same piece of ground annually. This
ground by being lightly manured in the
corn-hole every time it is planted, will produce corn for years, and give an ample supply.
When the first crop of corn is gathered in,
the corn stalks should be pulled up and
made to lie flat in the row where they grew,
so that they will not choke or impede the
growth of the neighbouring young corn.

In a former place I noticed the pressing
necessity, of taking proper care of what
ground or vegetable provisions, there might

happen to be on an estate. I now wish to lay before my readers the way to plant such provisions. I would begin with a patch for ground provisions. Supposing that three acres will be adequate to support the white people on an estate, I would look out for ten acres of as good land as could be found, adjacent to the cane culture, and in as sheltered a situation as possible, that was thrown out of such culture, by reason of its laborious distance for cane carriage, giving indifferent or bad sugar for want of proper exposure, or the soil too rank and spongy to continue in canes. This ten acres I would subdivide, first, in a patch of three acres for ground provisions, which I would parcel out into five or six divisions; the remaining seven acres should be laid out in a plantain walk, and be the most sheltered part of the ground. The whole of this ten acres of ground, should be well stocked up and hoed; if heavy in grass, or weeds, cleanly burned off, and any lodgments of water let off the land by a trench.

It should be fenced against trespass, by a double ditch topped with penguins, or a post and rail fence, on that side, which does not border on the cane cultivation.

Cocoas or eddoes are the most lasting and durable ground provisions, as they ratoon for years; and by planting them in successive patches, can be gathered in at any time of the year that they become ripe. There are several kinds of them; the bourbon, which is large, but rather soft, less nutritious, but more palatable to white people, and sooner becomes fit to dig in; and the country white and black cocoas, which are small, more prolific, more nourishing, drier, and more agreeable to the negroes, but which take a longer time to become ripe. I would appropriate two acres, out of the three that I intended for ground provisions, for planting and rearing cocoas, and the other acre for planting and rearing of yams. These two acres for cocoas, I would divide into four patches, in order that they may come in at different periods.

A patch should be planted every six weeks till the ground is occupied; two patches of bourbon, and two of country cocoas; the ground should be lined out two feet and a half square, and be pegged off; the gang should then be set in to dig the cocoa-holes at every peg, which should be six or eight inches deep, and a foot square. When the holes are opened, the bank must be hauled up to the edge of the holes and broke fine; then having obtained some good, ripe, dry cocoa heads, with good vegetating eyes, let them be cut into plants of two and a half inches square, with good eyes to each plant, and carried to the piece of ground. Each negro must take a parcel of these plants, and put two of them firmly in the bottom of each hole, three or four inches apart from each other, with the skin-side downwards, and cover them up well with the bank mold; and so finish the planting of the patch. At the end of six weeks, put in another patch; when this is planted, the former will be above ground, and may re-

quire cleaning and molding, which should
be attended to. In this manner proceed
in planting the two acres in cocoas, till
completed; taking care to clean and mold
the other patches which were planted. The
first patch will cover the ground, and be
thick strong stems, by the time the last is
planted. Cocoas will require two moldings
and three cleanings, before they begin to
button or bear at the roots, and get the
better, or overpower the growth of weeds
that spring up among them. Bourbon
cocoas will be ripe in nine months, and
country cocoas in twelve months after they
are planted. According as each patch is
dug, it should be well hoed; the cocoa
heads made steady in the ground, the old
leaves mostly lopped off the heads, (except
the top sucker) and the head well molded
up. After this, by giving them one molding, and two cleanings annually, there then
will be a succession of cocoas for years.

There are several kinds of yams. The
negro and white yams are principally cul-

tivated. They are the largest, most lasting, and nutritious; the Indian yam though floury, and delicate, is not much planted, is capricious and hard to be reared, and therefore not much in use. The yam is a yearly vegetable. The period for planting it is about Christmas for the negro yam, and February or March for the white yam. Care should be taken to have a sufficiency of yam heads or plants ready to be put in the ground, of good sound quality, against the proper time. The ground should be lined about four feet square, and pegged off to such distances. The gang must be set in with hoes to dig yam holes, or as it is termed, to raise yam hills; a negro to each peg in a fair breadth to dig the piece out. They are to form a circle of about 18 or 20 inches in diameter round the peg with the hoe, using the peg as a centre, (to which the mold is to be hauled up,) and dig the mold about three inches below the surface of the ground, within the circle. The mold is to be hauled up out of

the hole ; then the ground inside the hole
must be chopped to loosen it. After that is
done, the mold which has been dug out of
the hole, should be hauled up in a conical
form round the peg, the hole again filled up
with it, and some more mold scraped, or
lightly cut by the hoe from the space between the pegs, which is to be drawn, and
added to each conical hill; and thus the yam
holes dug, and the hills made high, bulky, and
pervious, for the reception of the yam head
or seed. When the yam heads are ready,
or prepared for planting, they should be of
good size, about half a pound weight each,
sound, and have lively eyes. Each negro
should then take a parcel of them, and
place two at each yam hill, till each hill
has its appointed plants stationed at it.
Then let each negro set in to plant them,
by cautiously introducing one of his hands
into the bosom of the hill, on one side of
it, make rather a deep opening for the
plant, and deposit it firmly, rather inclining
to the base of the hill, and the eye or vine

part of the plant uppermost. Then cover it up, and close the mold ; make an open-ing in like manner in the other side of the hill, introduce, and plant the other head in the same way, and in this manner, is the yam piece to be planted out till it is finished.

Corn may be planted through a yam piece immediately after it is planted with yams ; when the budding vine from the yam head, which has been planted in the hill, makes its appearance, peeping as it were from the side of the earth, a parcel of stiff prongy stakes should be obtained, and one drove firmly and perpendicularly down by the side of the hill, near where the vine is seen to shoot its tendril ; and when the creeping tendril, is long enough to bear coiling round the stake, a careful person should be made to help its efforts, and twine it gently and tenderly round the stake, that it may climb on, and branch from it.   This should be done at every hill as soon as the vine appears.   It assists the growth of the plant, which descends into the ground, as the vine advances in height and magnitude. The stakes being perpendicular above, the yam inclines to grow perpendicular downwards, and hinders it from appearing above the surface of the hill, which would give it a disagreeable, bitter flavour. The yam swells, lengthens, and descends into the ground as it grows. When the vine stops growing, or begins to wither, the yam is nearly ripe, and will be soon fit to dig. When grass or weeds begin to make them foul, they should be hoed off in the spaces, and the weeds or grass gently pulled from the hills. The hills should have additional mold, so that none of the yam may be uncovered, or appear above ground. In the month of August, one half of the negro yam piece may be cut for heads, that is, a number of negroes should be put in with sharp knives; and gently introducing one of their hands into the hill, where the yam is growing, with the other hand and knife, cut the head of the yam off, near the but of the vine ; then dig the bottom of the yam out of the hill or hole, leaving the but of the vine, with the small piece of the yam head on it, firmly sunk in the hill ; which, by Christmas, will again grow to a considerable large bulbous head, with a sizeable bottom to it, and will serve for future plants. The yams which have been dug out in this manner in August are young, they are to be brought home, and soon made use of for the table. They are similar to young potatoes at this time, being rather soft, moist, and will not keep long ; however, they are esteemed a rarity. At Christmas time, the negro yam vines are all withered; the whole of the yams should be dug in, carried home, cleared from earth, powdered with lime or ashes, then heaped up, so as not to bruise one another, and made use of when wanted. The white yams are to receive the same mode of management, only that they are not to be cut for heads or plants in August. They should be planted in February or

March, and gathered or dug in when their vines are dry, about February or March following. A piece of the head or body of the white yam, being left in the ground, will remain to bear for years there, only a little, or rather somewhat degenerated in size, but often dug up in great perfection.

I now come to that part of the system, which includes the laying out a plantain walk, which is much esteemed, for the perpetual, wholesome, nutritious supply of vegetable food which it affords the planter and the negro, if storms do not destroy it. The plot of seven acres which I reserved for its occupation, and which should be the most sheltered part of the ten acre piece of ground, I would have lined out ten feet by seven ; ten feet space, and seven feet distance. The plantain row must be pegged off every seven feet, and spaces of ten feet between row and row. The rows should be fair and parallel one with the other. This broad way of plant-

ing it is necessary, as the plantain root or
sucker is extremely prolific, the plantain
tree majestically thick, large and high,
when full grown, and many of these exuberant stems grow at once from the
same stock ; throwing out a foliage of sur-
prising shade and beauty. It is then necessary to plant them wide, in order to let them
have a freedom of air, and that the sun may
prevail to mature their fruit. The ground
being all cleared from grass, bushes, and
weeds, and lined out, the great gang should
be put in with hoes, to dig the plaintain
holes at every peg, a negro to each row.
The holes should be dug deep, two feet
long by sixteen inches broad, to give room
for the large, ponderous plantain sucker to
be placed in them. The mold must be
hauled up to the edge of the hole, and
broke if too large. The plantain suckers
being ready, and trimmed, each negro
should take some, and place one good
sucker at every hole in the piece, and begin to plant them, by taking a sucker, and

placing it with the but, or rooty end in the bottom of the hole ; make the sucker lie in a leaning, reclining, or half horizontal position in the hole, with the small, or sucker end of the plant a little above the ground ; and when thus placed, draw the mold from the bank, and cover the plant well with it, leaving a little of the plant above the ground. In this manner the plantain walk should be formed. In a few weeks (if the weather is favourable) the young plantain shoot will be seen rearing its perpendicular head, perhaps three or four growing from the same stock. They should then be carefully molded, and cleaned of grass and weeds, when they are a few inches high. No cavities or water logging holes should be near them. The banks must be levelled about them, the holes filled, and properly closed up, and some fine mold given them, to encourage their growth. There will be no occasion to give them more than two moldings till they are established, but they must

be carefully kept clear from weeds or
grass ; and when any dry trash happens to
be hanging about them, it should be gently
cut off with a knife, and placed about their
roots, to keep them either free from too
much sun or chill. A plantain walk, well
taken care of, will be in bearing twelve
months after it is planted, amply repaying
for the labour and trouble of planting it,
and giving an almost inexhaustible supply
of fine provisions, if the vicissitudes of
hurricanes or storms (which this climate is
unhappily subject to) does not destroy it,
and which no human foresight or care can
prevent. When a plantain walk is made,
there may be a row of cocoas in the middle
of the ten feet spaces, which will yield a
crop by the time the plantain walk bears
fruit, but they must then be pulled up. A
few banana suckers can be planted in the
plantain row, instead of plantain suckers ;
sometimes they are much in request, as a
luscious, wholesome fruit, and for the
strong, fine-flavoured vinegar which is pro-

duced from them ; after this piece of ground is thus planted, the whole of it may be sown through with corn, which will not injure the plantain suckers, or trees, if it is not too close or thick.

It may not be improper, but, I presume, serviceable, here to remark, that the valuable island of Jamaica affords a variety of indigenous medicinal plants, shrubs, and herbs, (besides many foreign plants, which have been imported into the colony, and grow as well as in their native soil and climate,) which are simple and efficacious in many disorders ; such as the physic, or oil-nut tree, the oil produced from the nut of which, when well boiled and clarified, is equal in value and virtue to the best prepared castor oil. The tree not only grows spontaneous here, but may be propagated in abundance in any spot, by depositing a couple of its ripe, sound nuts or seeds, in a small hole dug for the purpose.

The wangola, a small Indian shrub, whose leaf, when steeped for a short time in water, or simple liquid, makes it of a mucous, glutinous quality, and is deemed an excellent medicine in case of flux or dysentery. It may be propagated from the sucker or seed, and requires only care to keep it clean from weeds, and a sheltered situation to live and thrive in.

The Indian arrowroot, a plant of no mean character, in the opinion of the skilful part of the faculty, both as a medicine, in diseases of the bowels, and for its fine, delicate, nourishing qualities for convalescents. This plant can easily be generated and produced; it requires only to have a spot of good ground appointed for its jointed long root, or fibrous, long, sharp-leafed stem to be inserted or planted in. The ground must be prepared and dug, in a similar manner as that for a cocoa piece, the root or fibrous stem set in the holes when dug, in a reclining way, with a small part of the stem above ground. It soon takes root; and when it has struck, and fresh buds or leaves begin to shoot from it, they should

E E

be cleaned and molded with care, till the leaves cover the ground and begin to blossom. When the blossom falls off naturally, and the leaves wither, the roots are ripe, and may be dug up. They yield a perpetual annual crop, by leaving some of the fibrous stalks in the ground, and regularly molding and cleaning them. The flour or starch produced from the root (after it is washed) by compressure, and being steeped in pure water, is what is esteemed so much by the faculty.

The rhubard is quickly and luxuriantly produced here, by planting its bulbous root in the same manner as in English gardens. I need not expatiate on its virtues, as it is so generally known and made use of for various complaints, and even as a delicacy for the table.

The cotton tree, or shrub, is extremely useful in many cases; it gives its warm down, not only for many beautiful fabrications, but for surgical aid to the hospital, and supplying wick for lamps in crop time.

It is easily reared from the seed or sucker, and should be planted at seven feet distance from each other.

The palm-nut tree is an elegant and stately production of nature, whose luscious nut gives a fine flavoured, thick red oil, of a penetrating and invigorating kind, which can be made use of internally and externally, with simple, flexible, and healing effect, and without any dread or danger. It can be produced from the nut or sucker, planted at twelve feet distance from each other ; and when once established, is almost perpetual, and is ornamental in its appearance.

These especially, and several other plants of a healthful, salutary character, should be industriously sought after, and planted in a spot contiguous to the kitchen garden, on every property in the country parts of Jamaica, (as likewise the introduction of a few swarms of bees in such a place, which thrive well, and give delicious, fine-flavoured honey and wax,) as they are, and ever will be, by their nature and powers, of the utmost use, and important consequences, for the healthy, the sick, and convalescent of every degree. The planting, rearing, and production of them, by every proprietor or overseer of a property in Jamaica, will be highly meritorious, and lastingly beneficial.

THE END.

LONDON:
Printed by A. & R. Spottiswoode,
New-Street-Square.